Electric Power Ahead

Azhar ul H

Copyright

Copyright © 2025 by Azhar ul Haque Sario

All rights reserved. No part of this book may be reproduced in any manner whatsoever without written permission except in the case of brief quotations embodied in critical articles and reviews.
First Printing, 2025

Azhar.sario@hotmail.co.uk

ORCID: https://orcid.org/0009-0004-8629-830X
Disclaimer: This book is free from AI use. The cover was designed in Microsoft Publisher

Contents

Copyright 2
Unveiling the Electric Revolution and Solid-State Frontier 4
Titans of Global EV and SSB Innovation 14
Redefining EV Design Through Ingenuity 22
Crafting the Solid-State Future: Manufacturing Marvels .. 30
Economic Ripples of an Electric Shift 40
Greening the Grid: Sustainability in Focus 49
Tomorrow's Blueprint: Visionary Projects 57
The Consumer Pulse: Adoption Dynamics 66
Fortifying SSBs: Safety as Priority 74
Beyond Roads: SSB Versatility 84
Rules of the Road: Regulatory Realms 93
Minds at Work: R&D Ecosystems 102
Powering Up: Infrastructure Innovations 111
Culture in Charge: Societal Shifts 120
Trailblazers: Landmark EV and SSB Stories 128
Horizons Ahead: Challenges and Promises 136
The Road Forward: Vision and Victory 145
About Author 154

Unveiling the Electric Revolution and Solid-State Frontier

The Real Story of the Electric Car (It's Way Older Than You Think)

We think we know the electric car story. Tesla, Elon, maybe a golf cart lurking somewhere in the background. But the true story of electric cars? It's like finding a hidden room in a house you thought you knew. It's older, weirder, and way more interesting than the usual tale. We're talking way before Musk was even a gleam in anyone's eye. Forget the supposed "gasoline car always wins" narrative – the electric dream was buzzing to life almost two centuries ago.

Picture this: 1830s Hungary. Not exactly Silicon Valley, right? But you've got this Benedictine priest, Ányos Jedlik – a physicist, not a car guy – messing around with magnets and electricity. He wasn't trying to build the next Uber; he was just fascinated. Jedlik cooked up this tiny, almost toy-like electric motor. It was a little spark of genius that, believe it or not, is a direct ancestor of your neighbor's Tesla. He built a little model car. It moved. On electric power. Wrap your head around that for a second!

The plot thickens, naturally. It jumps oceans, skips decades. We tend to fast-forward to the gasoline engine's "big win," but electric cars were having their own little moments of glory. This isn't a straight line from A to B; it's more like a rambling country road with some fascinating detours.

Fast forward to the late 1800s. Cities were basically drowning in...well, horse poop. Seriously. Horse-drawn carriages were the norm, and they created a mountain of manure, attracting flies and a smell that would make you cry. Urban pollution wasn't just a buzzword; it was a full-blown health crisis. People were desperate for something cleaner.

Enter the Electrobat. No, seriously, that was its name – sounds like something Batman would reject, right? But this electric car, built in the 1890s by Henry G. Morris and Pedro G. Salom in Philadelphia, was surprisingly successful. They weren't sexy – imagine a clunky carriage without the horse – but that's exactly what they were meant to be. Believe it or not, some city post offices actually used them.

But, and this is a big "but," our pioneers hit the same wall we are still bumping into, batteries. The lead-acid batteries back then were like carrying around lead bricks, and you could forget about finding a charging station.

But don't write off the Electrobat as some historical footnote. These things actually worked in their own way. Short trips around town? Perfect. They were quiet, clean (no exhaust!), and didn't break down every five minutes like some of those early gas-guzzlers.

The IEEE (Institute of Electrical and Electronics Engineers) archives, if anyone bothers to dig through them, tell a very different story from the one we usually hear. The electric car wasn't some instant failure. It was a contender!

So, why did gasoline cars end up on top? Well, that's a whole other story – oil tycoons, Henry Ford's assembly line, and the open road's appeal, they, played a role. But it's vital to remember that the early days of cars weren't a knockout win for gasoline. Electric cars had a vibrant, often forgotten, start.

So, next time you see an electric car whiz by, don't just think of Elon Musk. Think of Ányos Jedlik and his little motor. Think of the Electrobat, hauling mail through a city desperate for clean air. The real genesis of electric mobility is a story of brilliant minds, desperate needs, and a very, very long road. A road that, after a few wrong turns, is finally circling back to where it all began.

Alright, buckle up, buttercup, because we're about to take a joyride into the electrifying world of solid-state batteries (SSBs)! Think of them as the rockstars of the battery world, poised to dethrone the current king, the lithium-ion battery. We're not just talking about a minor upgrade; we're talking about a potential revolution in how we power, well, everything.

Let's ditch the jargon for a second and get to the heart of the matter. Imagine your typical lithium-ion battery as a swimming pool. You've got the two ends (the electrodes), and the water (the liquid electrolyte) lets the little lithium-ion swimmers do laps between them. It works, sure, but it's a bit messy, a little temperamental, and sometimes those swimmers get a bit… unruly (we'll get to that).

Now, picture an SSB. Instead of a pool, we've got a super-sleek, high-tech, solid-state trampoline park for ions. No sloshing, no leaks, just pure, solid-state awesomeness. This is where the magic happens.

1. The Solid Electrolyte: The Heartbeat of the Revolution

Forget that flammable, sometimes-leaky liquid electrolyte. The SSB's core is a solid, unwavering champion – the solid electrolyte. It's the meticulously designed track that allows our lithium-ion athletes to sprint back and forth, powering your gadgets with electrifying speed.

One of the rockstars of this solid-state world is a material called LLZO (Lithium Lanthanum Zirconium Oxide). Sounds like something out of a sci-fi movie, right? It's a type of ceramic, but not the kind you'd find in your grandma's china cabinet. This is engineered ceramic, a high-performance material crafted to be an ionic superhighway. Think of it as a perfectly paved road with dedicated express lanes for lithium ions.

Why is this better than the "swimming pool"? Because these ceramics are incredibly stable, even when things get hot. They're like the stoic, unflappable bouncers of the battery world, keeping everything in order and preventing any unwanted shenanigans (like those pesky dendrites – more on them in a bit!).

2. The Lithium-Metal Anode: Unleashing the Beast

Now, let's talk about the powerhouse of the SSB – the anode. In your everyday lithium-ion battery, the anode is usually made of graphite. Think of graphite as a reliable, but somewhat modest, storage unit for lithium.

SSBs, however, are aiming for something much bolder: a lithium-metal anode. This is like switching from a small storage locker to a massive, state-of-the-art warehouse. Lithium metal can hold way more lithium, which translates to a battery that's smaller, lighter, and packs a much bigger punch. We're talking about potentially doubling or even tripling the energy density!

But, there's a catch. Lithium metal is a bit of a diva. It's highly reactive and, frankly, a bit of a troublemaker. It's super prone to forming those dendrites – think of them as tiny, metallic stalactites that grow inside the battery and can cause short circuits. It's like the warehouse is so eager to store stuff that it starts piling things up haphazardly, creating a fire hazard. So, scientists are working hard to tame this wild child, using clever tricks to keep the lithium metal in line.

The Ionic Conductivity Game: It's All About the Flow

The key to a great SSB is how smoothly those lithium ions can move through the solid electrolyte. This is called ionic conductivity, and it's like the traffic flow on our ionic superhighway.

Liquid electrolytes are generally pretty good at this – they're like wide-open freeways. But, as we've discussed, they have their downsides. Getting a solid material to match that level of flow is a real challenge. Imagine trying to navigate a dense forest versus a clear path. The crystalline structure of ceramics can be like that forest, making it harder for ions to find their way.

But, researchers are like expert trail blazers, constantly finding new ways to make the journey easier. They're experimenting with different recipes for the solid electrolyte, adding secret ingredients (doping), and even redesigning the "forest" itself (the microstructure) to create clear pathways for the ions. It is the material equivalent of landscape design.

Real-World Heroes: QuantumScape and the Path to Progress

This isn't just some futuristic fantasy. Companies like QuantumScape are out there in the trenches, building and testing these batteries right now. They're like the pioneers, charting the unknown territory of SSB technology. They're sharing their discoveries, showing us how their solid electrolytes perform, and how they're tackling those pesky dendrites. It's like watching a live-action science experiment, and the results are getting more and more exciting.

In a nutshell, the solid-state battery is a beautiful blend of cutting-edge science and engineering wizardry. It's about taking the battery, a device we rely on every day, and completely reimagining it from the ground up. It's a bold, ambitious project, but the potential payoff – safer, longer-lasting, more powerful batteries – is worth the effort. It's not just about better gadgets; it's about a potentially greener, more sustainable future. So, keep your eyes peeled, because the SSB revolution is just getting started!

Solid-State Batteries: The Electric Dream (and the Microscopic Hurdles)

Solid-state batteries. Say it out loud – it sounds like the future, right? We're talking about a potential game-changer, a technology that could make our current batteries look like, well, relics. Imagine your phone lasting for days, your electric car cruising for hundreds of miles extra, all thanks to batteries that could boast 50% more energy density than today's lithium-ion champions. That's like going from dial-up internet to fiber optic – a total leap.

But hold on a second. Before we all get carried away with visions of endless power, there's a catch (isn't there always?). It's like this: we've got the blueprint for an amazing skyscraper, but we're still figuring out how to make the concrete strong enough.

The main villain in this story? Something called interfacial resistance. It's a mouthful, but the concept is pretty simple.

Picture a battery as a delicious sandwich. The electrodes are the bread, and the electrolyte is the tasty filling that lets the electricity flow. In current lithium-ion batteries, that filling is a liquid – everything's nicely connected. But in a solid-state battery, the electrolyte is solid. Imagine that sandwich again, but this time the filling is a solid chunk of, say, hard cheese. Getting that cheese to really connect with the bread on every single bit of surface? That's the problem.

This interfacial resistance is like throwing sand in the gears. It makes it hard for the ions (the tiny electricity carriers) to move around. This means slower charging, weaker performance, and, worst of all, the potential for dendrites – those nasty little metallic spikes that can short-circuit a battery and cause a, shall we say, fiery situation.

Publications like the Journal of Power Sources are filled with scientists wrestling with this. They're like microscopic architects, trying to design the perfect interface. They're trying everything. New materials. Different pressures. Think of squeezing the battery super tight while making it. They are trying to find some magic combination.
It isn't something only discussed in labs. Even Toyota, a company that's been betting big on solid-state batteries, has been honest about the challenges. They've made progress, sure, but scaling up production, making sure every battery works perfectly, and ensuring they last for years – that's still a work in progress.

The lure of all that extra power keeps everyone going. Scientists are playing with all sorts of solid electrolytes: ceramics that can withstand high temperatures, flexible polymers, even glass-like materials. Each one is like a different ingredient in a recipe, and they're still searching for the perfect blend.

So, are solid-state batteries just a pipe dream? Absolutely not. They're real, they're being tested, and they hold incredible potential. But they're like teenagers – full of promise, but still going through some growing pains. Getting them to the point where they can power our world? That's going to take more time, more research, and a whole lot of ingenuity. It's a scientific marathon, not a sprint. But the finish line? A future with safer, longer-lasting, and dramatically more powerful batteries. It's a future worth fighting for, a future that could reshape how we live, work, and travel. It is a big deal.

The Silent Revolution: Your Future Garage Will Be Seriously Charged (But in a Good Way)

Remember that nervous twitch you get when your phone battery dips below 20%? That's nothing compared to the cold sweat of range anxiety, the EV owner's arch-nemesis. For years, electric cars have been like that brilliant, quirky friend who's almost perfect, but has that one annoying habit (running out of juice) that keeps you from fully committing. But what if that habit vanished? What if we could ditch the charger-hunting anxiety and embrace a future where "miles to empty" becomes a quaint, forgotten phrase?

Enter the solid-state battery (SSB) – not some futuristic fantasy, but the very real technology that's about to kick range anxiety to the curb. Think of it like this: your current EV battery is like a water balloon – powerful, but a bit...

precarious. The liquid sloshing around inside (the electrolyte) is essential, but it's also flammable and can get a little unstable. SSBs, on the other hand, are more like a solid, dependable brick. They swap that liquid for a solid electrolyte, and that seemingly small change unleashes a torrent of benefits.

From NYC to Detroit on a Single Charge? Seriously?

We're not talking about baby steps here. We're talking about giant, moon-landing-sized leaps in range. Forget nervously eyeing the battery gauge every few miles. Companies like Solid Power, who are teaming up with BMW, aren't messing around. They're talking about EVs that can cruise for over 600 miles on a single charge. Imagine that road trip: New York to Detroit, windows down, music blasting, and zero worries about finding a charging station in the middle of nowhere. This isn't science fiction; it's the very near future, backed by real-world testing and some seriously impressive materials science (think swapping out graphite for energy-packed lithium metal – it's like upgrading from a AA battery to a power plant).

Safety First (Because Nobody Wants a Spontaneous Car-B-Que)

Let's be honest, those occasional news stories about EV fires haven't exactly been reassuring. The liquid electrolyte in current batteries can be a bit of a fire hazard. SSBs, however, are the cool-headed cousins of the battery world. Their solid electrolyte is inherently more stable, drastically reducing the risk of those dramatic thermal runaway events. It's like trading a Molotov cocktail for a really sturdy paperweight. You're getting the power, but without the potential for explosive surprises.

Mercedes-Benz: Putting the "Solid" in Solid-State

Daimler (the folks behind Mercedes-Benz) aren't just sitting on the sidelines; they're diving headfirst into the SSB revolution. They're running pilot projects, pushing these batteries to their limits, subjecting them to all sorts of real-world torture tests. It's not just about proving that SSBs can work; it's about making them work flawlessly, reliably, and for years to come. They're essentially building the foundation for the next generation of luxury EVs – ones that are as smooth and safe as they are powerful.

The Road Ahead: Still Bumpy, But Paved with Promise

Okay, let's not pretend this is all happening tomorrow. There are still some hurdles. Making these batteries affordable and mass-producible is a challenge, like trying to build a skyscraper out of LEGOs – it's doable, but it takes some serious engineering.

But the momentum is undeniable. The pioneers, the dreamers, the engineers at companies like Solid Power, BMW, and Daimler are laying the groundwork for a truly electric future. Your future garage might not have the roar of a gasoline engine, but it'll have something far more exciting: the silent hum of an EV that can go farther, safer, and with more confidence than ever before. Get ready for the silent revolution. It's going to be electrifying.

Titans of Global EV and SSB Innovation

China's EV Revolution: More Than Just Sparks Flying

Forget everything you thought you knew about slow, steady progress. China's electric vehicle (EV) story isn't a gentle evolution; it's a full-blown revolution, a lightning strike that's reshaping the global car industry. They're not just building cars; they're building an electric empire, and they're doing it at warp speed.

Imagine a grand, ambitious plan – not some dry, bureaucratic document, but a vibrant blueprint for the future. That's China's 14th Five-Year Plan. It's like a conductor's score for an electric orchestra, calling for a massive crescendo of EVs. The message is loud and clear: "Go electric, or go home." This means showering EV makers with support, blanketing the country with charging stations, and making it increasingly painful to stick with old-fashioned gas guzzlers.

But a plan is just paper without the muscle to back it up. Enter the titans of industry. Think of CATL, the battery behemoth. This isn't just a battery company; it's the battery company. It's like the Willy Wonka's chocolate factory of EV batteries. Reportedly, around half the world's EV batteries are flowing from this single Chinese powerhouse. That's not just market share; that's near-total domination of a critical component. It happened because of backing from the governement.

And CATL isn't alone. Companies like BYD, NIO, and Xpeng are like a pack of electric wolves, hungry for innovation. They're churning out EVs that are not only cheaper but are also increasingly packed with cool tech and features, giving established carmakers a serious run for their money.

They're building entire ecosystems – charging networks, battery-swapping stations that feel like futuristic pit stops, and digital services that weave seamlessly into your daily life.

Want to see this revolution in action? Take a trip to Shenzhen. This megacity is like a glimpse into the future. Every single bus is electric. Imagine that: a city of millions, moving silently and cleanly on electric power. It's not a sci-fi fantasy; it's real, thanks to Shenzhen's commitment. BloombergNEF's data backs this up – it's proof that a fully electric public transport system isn't just a pipe dream. It's happening now, and it's spreading like wildfire across China.

Shenzhen's story is showcasing what a goverment with clear targets and comitted companies with prodcution power can do.
The commitment and capabilities are working like a well-oiled machine.

China's EV dominance is about far more than just selling cars. It's about controlling the future of how we move. They've essentially planted their flag on the electric mountain, becoming the undisputed leader in a technology that will redefine how we power our lives, fuel our economies, and even interact on a global scale. While the rest of the world is still arguing about how fast to go electric, China is already miles down the road, building the highways, the power plants, and the supply chains of tomorrow. It's bold, it's breathtaking, and frankly, it's a little intimidating. The world is now in a race to catch up, and the starting gun fired a long time ago.

America's Battery Moonshot: Bendy Power is the Future

Forget the rockets for a sec. The real space race is happening right here on Earth, in labs and factories where the next energy revolution is brewing. We're talking batteries, baby – but not the leaky, fire-prone kind in your grandpa's flashlight. We're talking solid-state batteries, and America's betting big on a secret weapon: polymers. Think bendy, stretchy, super-safe power.

Imagine MIT, not as some stuffy academic ivory tower, but as a battery bootcamp. Picture brainy folks in lab coats, not just tinkering, but wrestling with the very molecules that will power our future. They're like alchemists, but instead of turning lead into gold, they're turning squishy polymers into the heart of a battery that won't explode if you accidentally sit on it. That's the dream, anyway, and they're closer than you think.

But this ain't just about beakers and whiteboards. Enter the rebels: startups like Ionic Materials. These aren't your hoodie-wearing, app-developing Silicon Valley types. These are the material science mavericks, the dendrite destroyers. Dendrites? Those are the microscopic gremlins that can turn a battery into a tiny bomb. Ionic Materials says they've got a polymer that acts like a bouncer at a club, keeping those dendrites out. If they're right, it's a game-changer. They are selling the "secret sauce".

And who's buying? Think Ford, but instead of just making cars, they're placing bets like a seasoned gambler at a high-stakes poker game. They're spreading their chips across the table, investing in multiple solid-state technologies, with polymers being a major player. They're not just dreaming of electric cars; they're dreaming of electric cars with batteries that can handle a beating and still keep going. And it's all backed up by real science – the kind you find in journals with names like ACS Energy Letters, where eggheads publish the proof that this stuff actually works.

The cool thing about polymers? They're the chameleons of the battery world. They're not brittle and fussy like some of the other solid-state contenders. They're flexible, meaning we can potentially use the same machines that make today's batteries to crank out these new, supercharged versions. That means faster, cheaper, and maybe even... wearable batteries? Think power suits, but less Iron Man, more... Iron Pajamas.

Okay, okay, it's not all sunshine and roses. Getting these polymer batteries to conduct electricity as well as the liquid stuff is like teaching a sloth to run a marathon – it's tough. And making enough of them to power all those shiny new electric cars? That's a whole other mountain to climb.

But here's the thing: America's not backing down. It's not about one genius in a garage; it's a whole team effort. Universities, startups, and even the big guys are all in this together, chasing the same holy grail: a battery that's safer, lasts longer, and won't turn your pocket into a furnace. And the bendy, stretchy, polymer-powered path? It's looking mighty bright. The future of energy might just be surprisingly flexible.

Europe's Going All-In on Electric: It's More Than Just a Dream

Okay, picture this: Europe isn't just talking about going green; they're putting their money where their mouth is – and we're talking serious money. Think €100 billion. That's not pocket change; that's a full-blown commitment from the European Union to make electric vehicles (EVs) the kings of the road. This isn't some fluffy, "someday" promise; it's a full-throttle plan to ditch gas guzzlers and hit carbon neutral by 2045.

Carbon neutral. Let that sink in. It's a HUGE goal. But instead of just wishing on a star, the EU is rolling up its sleeves and getting to work. They are using this money, and it is not just for show. The European Commission's reports lay it all out: they're making EVs cheaper, boosting battery tech, and building a charging network that can handle a ton of electric cars. It is a team effort, crossing borders, for the future of the enviorment.

Want proof it's working? Check out Germany. They've made EVs way more affordable with some serious subsidies. And guess what? People are actually buying them! It's a no-brainer, really. Make electric cars cheaper, and people will drive them. It's not rocket science; it's common sense, backed by cold, hard cash.

But Europe's not just thinking about today; they're playing the long game. They don't want to trade one dependency (foreign oil) for another (foreign batteries). Enter Northvolt, the Swedish battery rockstars.

Northvolt is like the cool, innovative kid on the block, and they're building a "gigafactory" – think Willy Wonka's chocolate factory, but for batteries – that's all about sustainable power. They're laser-focused on solid-state battery (SSB) technology, and that's a big deal.

Why? Because SSBs are like the superheroes of batteries. Imagine a battery that's safer, packs more punch, and charges way faster than what's in your phone or EV right now. That's the SSB promise. We're talking longer drives, shorter pit stops, and a safer ride overall. Northvolt is basically at the cutting edge of battery tech, not just in Europe, but worldwide.

This Northvolt factory? It's more than just a building. It's a symbol. It's Europe saying, "We're not just playing in the EV game; we're here to win it." It's about creating a greener future, controlling their own EV destiny, and, let's be honest, creating a ton of jobs in the process.
The funds provided by the EU, the ingenuity of a company like Northvolt; what a powerful duo.
Europe isn't just joining the EV party; it's throwing the party. The road to 2045 is long, but with this kind of momentum, Europe's not just driving in the right direction; it's flooring it. The future? Buckle up, because it's electric.

Beyond Silicon Valley: Asia's Electric Dreams are Charging Ahead

Forget, for a moment, the gleaming Teslas of California. The real electric revolution is humming to life in the bustling cities and innovative labs of Asia. It's a revolution powered not just by billion-dollar valuations, but by a fierce determination to solve real-world problems, one battery, one electric vehicle, at a time.

Let's journey to South Korea, where a company called SK Innovation is quietly playing alchemist. They're not just tinkering; they're chasing the philosopher's stone of the battery world: solid-state batteries (SSBs). Imagine a phone that charges in minutes and lasts for days, or an electric car that can cross a country on a single charge, without a hint of range anxiety. That's the promise of SSBs, and SK Innovation is betting big on making it a reality. They're not just suppliers; they're dream-weavers, crafting the power source for a cleaner, more efficient future.

Then, hop over to India, where the rumble of Tata Motors is becoming a symphony of change. Tata, a familiar name in India, is playing a crucial, yet often overlooked, role on the global stage. They're not chasing the luxury EV market; they're building electric cars for the people. The Tata Nexon EV isn't just a car; it's a statement: electric mobility shouldn't be a privilege, it should be a right. It's about empowering millions to embrace cleaner transport, and in doing so, addressing the very air they breathe. It's democratization, powered by electricity.

But this isn't just about blueprints and boardrooms. It's about seeing the future unfold. Picture Seoul, a city that breathes innovation. Here, electric taxis, often powered by the very batteries SK Innovation is perfecting, aren't just ferrying passengers; they're gathering wisdom. They're mobile research labs, silently collecting data on battery life, charging needs, and even how drivers and riders feel about this new way of moving. Seoul isn't just adopting electric vehicles; it's living the electric future, learning and adapting in real-time.
It is a leap of faith, with the future in mind.
Seoul is showing commitment, and providing reasurch, for all.

These aren't just companies; they're solutionaries. They're tackling the giants of our time: the choking smog in our cities, the looming shadow of climate change, and the need for energy independence. They're showing the world that innovation isn't confined to Silicon Valley; it can sprout from the most unexpected corners of the globe. They are empowering the people, to show that they too can do great things.

So, the next time you think about the future of electric mobility, look East. Look beyond the headlines and the hype. The real story is being written by the unsung heroes,

the quiet revolutionaries, the companies like SK Innovation and Tata Motors, who are proving that a cleaner, more sustainable future is not just possible, but is already being built, one electric dream at a time. They are writing the future, with sustainable ink.
They are the true champions of the future.

Redefining EV Design Through Ingenuity

The EV Diet Plan: Shedding Pounds with Super Materials

Let's be honest, nobody wants a heavy EV. It's like running a marathon in hiking boots – you're not going to get very far, very fast. The electric car revolution isn't just about ditching the gas pump; it's about a total vehicle makeover, and that includes a serious weight-loss program.

We're not talking about crash diets here. We're talking about Material Alchemy – the almost magical process of creating materials that are ridiculously light and ridiculously strong. It's like giving cars a superhero upgrade.

Two materials are basically the rockstars of this lightweight revolution:

    Graphene Composites: The Spider-Silk of Cars: Imagine something as thin as a single layer of atoms, but stronger than steel. That's graphene. Now, picture weaving that into other materials like a super-strong spider web. That's a graphene composite. Scientists (think of them as modern-day alchemists) are figuring out how to use this stuff in everything from the car's "skin" (body panels) to the battery's home (the casing). The result? Lighter cars that can zoom further and handle like a dream.

    Magnesium Alloys: The Aluminum That Hit the Gym: We all know aluminum is lighter than steel. Well, magnesium is like aluminum's even lighter cousin, and when you mix it with other metals (that's the "alloy" part), it gets seriously strong. Think of it as aluminum that's been hitting the gym hard. Lotus, the folks who build those ridiculously fast sports cars, are masters of this. Their Evija electric hypercar is shockingly light for its power, thanks in part to magnesium.

But This Isn't Just for Supercars...

This isn't some exclusive technology for millionaires. Companies like Rivian are using clever material tricks to keep their electric SUVs light and efficient. And that's a win for everyone. Why?

    Smaller Battery = Cheaper Car: A lighter car doesn't need a massive, expensive battery to go the distance.
    Faster, Nimbler, More Fun: Less weight means quicker acceleration and sharper handling. It just feels better to drive.
    Real-World Range: You actually get the range the manufacturer promises (or close to it!).

The Balancing Act: Light, Strong, Safe, and Affordable

It's a juggling act, for sure. These materials need to be:

    Light: Obviously!
    Strong: To handle everyday bumps and scrapes.
    Safe: To protect you in a crash.
    Affordable: So we can all actually buy these cars.

And it is not only the materials but, the actual construction. Using different material thickness, adding voids like a honey comb structure, or gas pockets.

Graphene, for example, is still pretty pricey, but that's changing fast. The real challenge is finding that "Goldilocks" combination – the perfect materials and manufacturing methods that give us everything we want without breaking the bank.

It's like a giant, super-important puzzle, and the pieces are made of cutting-edge materials science. The companies that solve this puzzle will be the ones leading the charge in the electric car revolution. It's a race to build the lightest, most efficient EVs, and it's all thanks to the magic of material alchemy.

Dancing with the Wind: Reimagining How We Move

Ever stuck your hand out of a car window on the highway? That pushback, that invisible hand holding you back? That's air – and it's more powerful than most of us realize. For ages, we've treated air resistance like an annoying older sibling, always slowing us down, costing us fuel, and generally making things harder. But what if we could change that relationship? What if, instead of a wrestling match, we could turn it into a dance?

That's the revolution happening right now in aerodynamic design. Forget tiny tweaks and incremental gains. We're talking about a fundamental shift in thinking – a complete reimagining of how things move through the air. It's about becoming artists, not just engineers. We're learning to sculpt the air itself, coaxing it to flow around our vehicles like a gentle stream around a smooth stone.

Look at the Aptera. It's like a spaceship landed in our time, and that's no accident. Its wild, beautiful curves aren't just for looks. They're a masterclass in cheating the wind. With a drag coefficient (that's the fancy term for how slippery something is in the air) of just 0.13, the Aptera whispers through the atmosphere. Compare that to your average car (0.25 to 0.35) – it's like the difference between wading through molasses and gliding on ice. That difference? It could mean a whopping 30% boost in efficiency. Imagine longer road trips on a single charge, or filling up your gas tank way less often. The covered wheels? The details show an artists care.

The Aptera's secret? It's borrowing from nature's playbook. Think of the perfect teardrop shape – the way a raindrop falls, or the curve of a bird's wing in flight. Every inch of the Aptera is designed to minimize the chaotic swirls and eddies of air (turbulence) that create drag.

And it's not just the Aptera leading the charge. The Lucid Air, another electric beauty, can travel an astonishing 520 miles on a single charge. Sure, batteries are important, but aerodynamics are its secret weapon. The brainy folks at the American Institute of Aeronautics and Astronautics (AIAA) have proven, time and again, that even tiny changes in shape can have huge impacts on performance. The Lucid Air, with its smooth lines and carefully crafted underbelly, is proof that this isn't just lab theory – it's real-world magic.

But this "air sculpting" isn't just for futuristic EVs. Imagine semi-trucks that slice through the wind, saving mountains of fuel and reducing pollution. Picture bullet trains that practically fly along the tracks, reaching incredible speeds with less energy. Even airplanes could learn a thing or two.

The future of movement isn't about bigger engines or more power; it's about finesse. It's about understanding the subtle, powerful dance between an object and the air it moves through. We're leaving behind the era of brute force and stepping into a world of elegant, efficient motion. It's a future where we become partners with the wind, shaping it to our will, creating a smoother, greener, and honestly, more beautiful way to travel. It's a future of flow, and it's not just exciting – it's happening now.

Brainy Batteries: Giving Power a Personality

Forget everything you thought you knew about batteries. We're not just talking about swapping out liquids for solids anymore. Solid-state batteries (SSBs) are coming, and they're bringing a whole new level of smart with them. It's like going from a flip phone to a supercomputer – the difference is that dramatic. And the secret weapon? It's all in the brains.

Imagine a battery that doesn't just store energy, but understands it. That's the promise of the next-generation Battery Management System (BMS), the unsung hero in the SSB revolution. Think of the old BMS as a diligent hall monitor: making sure things don't get out of hand, preventing meltdowns (literally). Important, yes, but a bit...reactive.

The new BMS, powered by machine learning, is more like a world-class personal trainer for your battery. It's not just watching for problems; it's constantly learning, adapting, and optimizing. It's like having a tiny, incredibly intelligent energy guru living inside your battery pack.

This isn't your grandpa's battery management. This is AI-powered performance enhancement. Instead of just saying "don't go over this voltage," the new BMS is asking, "How can we squeeze every last drop of performance out of this system, safely and efficiently?"

General Motors is betting big on this with their Ultium platform. They're not just building batteries; they're building intelligent powerhouses. Their BMS is like a master conductor, orchestrating every aspect of the battery's operation. It's constantly tweaking and tuning, learning from every charge and discharge, predicting how the battery will behave in different conditions. It is like the BMS is saying, "Okay, you're driving up a mountain? Let me adjust for that. Now you're coasting downhill? I'll optimize for energy recapture."

It is thanks to a digital twin. The computer makes a virtual copy of the battery, constantly fed with real-world data. This allows to run "what if" scenarios, predicting how the real battery will react before anything even happens. It's like having a crystal ball, but powered by science, not magic.

And this isn't just some futuristic fantasy. Look at cars like the Hyundai Ioniq 5. While it's not fully solid-state yet, it's a sneak peek at what's coming. It's already learning from your driving habits, adjusting to the weather, and making sure you get the most out of every charge. It's a taste of the future, where your battery is almost a living, breathing part of your car.

The consequences are huge. These aren't just batteries that last a bit longer or charge a bit faster. We're talking about a fundamental shift in how we think about energy storage. Longer lifespans, faster charging, and increased safety are just the beginning. From powering our homes to revolutionizing transportation, "brainy" batteries are about to make a serious impact. These aren't just batteries; they're power with personality. Get ready for the smart energy revolution.

The Sleepless City: When Robot Cars Get a Superpower (And It's Not What You Think)

We've all heard the electric car hype – greener, cleaner, yadda yadda. But hold on a sec. What happens when you throw self-driving into the mix? Buckle up, because it's about to get weird (in a good way). We're not just talking about a Sunday drive anymore; we're talking about robots that never need a nap.

See, a self-driving car is basically a super-smart Roomba with a bigger travel budget. And just like your Roomba needs to dock and recharge, robot cars need juice. Traditional batteries? They're like that friend who always needs to stop for coffee – a little draining, right? They work, sure, but they limit how long these robotic road warriors can stay on the job. Downtime for a delivery bot, a futuristic taxi like the Cruise Origin, or even Baidu's Apollo Go robots rolling through city streets, means lost money, missed deliveries, and generally grumpy customers (even if those customers are other robots).

Enter the unsung hero: the solid-state battery (SSB). Think of it as the super-soldier serum for electric vehicles. SSBs are packed with energy, meaning our robot friends can go way longer without needing a pit stop. We're not talking about a slightly longer coffee break; we're talking about potentially days of non-stop operation. Imagine a city where delivery vans zip around 24/7, restocking shelves, delivering packages, and generally making life easier, all without ever needing to "sleep."

This isn't just science fiction. The brainiacs at places like the Robotics Institute (yeah, those Carnegie Mellon folks are at it again) are already cooking up this future. They're not just building smarter robots; they're building robots that are practically indestructible, at least when it comes to power. They're publishing papers – real, serious scientific stuff – about how robots, amazing batteries, and the real world can all play nicely together.

Take the Cruise Origin. It's not your grandma's Buick. It's a rolling platform, a blank canvas for whatever autonomous dreams we can conjure up – ride-sharing, package delivery, maybe even mobile pizza ovens (one can dream!). Now, imagine that Origin, supercharged with an SSB. Suddenly, it's not just a car; it's a tireless workhorse, maximizing every minute, optimizing every route, and basically running the city while we humans are catching some Z's.

And Baidu's Apollo Go? It's like a live-action experiment, a glimpse into this always-on future. It's showing us how these robots handle the chaos of real-world traffic, collecting data that'll be crucial when we swap out those old batteries for the super-powered SSBs.

This isn't just about building a better car; it's about building a better system. It's about a future where logistics are seamless, where transportation is hyper-efficient, and where robots can finally ditch the charging cable and truly roam free. It's about cities that never sleep, powered by robots that never tire. It's a little bit wild, a little bit futuristic, and a whole lot exciting.

Crafting the Solid-State Future: Manufacturing Marvels

Solid-State Batteries: Building the Future, One Atomic Layer at a Time (and Why It's Like the Perfect Soufflé)

Forget clunky, leaky batteries. The future is solid, and it's electric. We're talking about solid-state batteries (SSBs) – the superheroes of the energy storage world. They promise to power our electric vehicles for epic road trips, keep our phones alive for days, and make renewable energy truly reliable. But how do we actually build these powerhouses? It's all about the ingredients, and the chef.

Think of an SSB like a perfectly crafted soufflé. It needs to be light, airy (in a metaphorical, ionic-conductivity sense!), and flawlessly structured. The key ingredient? The solid electrolyte, specifically the sulfide-based ones. This is the "batter" of our battery, the medium through which lithium ions – our tiny, energetic delivery crew – race back and forth, carrying the charge.

Now, the old way of making this "batter" was a bit…messy. Imagine trying to make a soufflé with lumpy flour and uneven heat. You'd end up with a dense, uneven mess. Traditional methods of making sulfide electrolytes often resulted in similar problems: microscopic "lumps" (grain boundaries) and structural imperfections that slowed down our lithium-ion delivery crew. The result? A battery that was, well, a bit of a flop.

Enter the master chef: Vapor Deposition. This isn't some sci-fi magic trick, although it feels like it. Think of it as the ultimate in controlled cooking. Instead of dumping all the ingredients in at once, vapor deposition is like gently misting the perfect blend of sulfur, lithium, and other secret spices (phosphorus, germanium – shhh!) onto a surface. It's like airbrushing with atoms, building up the electrolyte layer by layer, with exquisite precision.

Why is this so important? Because this level of control creates a perfect structure. No more roadblocks for our lithium-ion couriers! They can zoom through the electrolyte with incredible speed and efficiency. We're talking about a superhighway for ions, leading to a battery that charges faster, lasts longer, and delivers more power.

And the proof is in the pudding (or, uh, soufflé). Real research – the kind published in fancy journals like Chemistry of Materials – shows that vapor deposition can boost ionic conductivity by a whopping 20%, and sometimes even more. That's not just a tiny improvement; it's a game-changer. It's the difference between a phone that dies halfway through the day and one that keeps going (and going, and going).

This isn't just lab-coat fantasy, either. Companies like SES are already whipping up batches of these batteries in their pilot lines, proving that this isn't just a theoretical recipe – it's ready for the real world. They are proving that the process can scale.

The ability to create these ultra-thin, perfectly uniform electrolyte layers isn't just about making better batteries. It's about making different batteries. Imagine batteries so thin and flexible they can be woven into your clothes, or batteries that can power a Mars rover in the freezing depths of space. Vapor deposition is unlocking a whole new cookbook of battery designs.

So, the next time you think about the future of energy, remember the humble sulfide electrolyte and the magic of vapor deposition. It's a testament to how materials science, with a dash of creativity and a whole lot of precision, is building a brighter, more powerfully charged tomorrow. It's all about mastering the materials, one perfectly placed atom at a time. And maybe, just maybe, enjoying a delicious soufflé along the way.

From "Sandwich Slapping" to Surgical Precision

Building an SSB cell isn't your weekend DIY project. It's more like performing surgery on a butterfly's wing – in a vacuum-sealed bubble. We're dealing with layers thinner than a human hair, tolerances tighter than your favorite skinny jeans after Thanksgiving, and an obsession with cleanliness that would make a germaphobe proud.

The core of the SSB is the "stack" – think of it like a layered lasagna of electrochemical goodness. But instead of delicious pasta and cheese, we're talking about highly reactive materials that hate moisture and dust more than a vampire hates garlic. A single stray dust particle? A microscopic misalignment? That's like throwing a wrench into a finely tuned Swiss watch. It can lead to anything from a slightly underperforming battery to, well, let's just say "fireworks" are not the desired outcome.

Laser Beams: Not Just for Sci-Fi Anymore

Forget clunky soldering – we're wielding the power of light! Laser welding is the key to joining these delicate layers. Imagine a surgeon using a laser scalpel, but instead of cutting, they're fusing materials together with pinpoint accuracy. It's like microscopic spot welding, creating incredibly strong bonds without turning the whole thing into a molten mess.

The Journal of Manufacturing Processes – yeah, it's a real thing, and it's surprisingly exciting when you're talking about lasers – dives deep into the nitty-gritty. It's all about finding the Goldilocks zone: enough laser power to melt the materials just right, but not so much that you vaporize your precious electrolyte. Pulse duration, scanning speed... it's a delicate ballet of physics, and getting it wrong means a battery that's more fizzle than sizzle.

The Dry Room: Where Humidity Goes to Die

Now, picture this: a room so dry, it makes the Atacama Desert look like a rainforest. That's the "dry room" where SSB assembly takes place. We're not talking "slightly less humid," we're talking "your-lip-balm-will-become-your-best-friend" dry.

Why the extreme dryness? Because many solid-state electrolytes are basically moisture magnets, but in a bad way. They react with water vapor, creating gunk that messes with the battery's performance. So, we're talking spaceships and the international space station, but on Earth.

These dry rooms are like cleanrooms on a superhero-level dose of cleanliness. Filtered air, full-body bunny suits (think less Easter, more hazmat), and equipment designed to shed fewer particles than a bald eagle in winter. It's borderline obsessive, but absolutely crucial.

CATL: The Battery Ninjas

Companies like CATL, the Chinese battery behemoth, aren't just following the rules; they're rewriting them. They're like the battery ninjas, constantly refining their techniques, pushing the limits of what's possible.

They've got real-time monitoring systems that are like the hawk-eyes of the manufacturing world. These sensors can detect the tiniest wobble during welding, allowing for on-the-fly adjustments. It's all about consistency, consistency, consistency. One tiny slip-up, and you've got a batch of batteries that are more likely to power a paperweight than an electric car.

Their secret sauce? It's a mix of relentless innovation, a willingness to push boundaries, and probably a whole lot of caffeine. They're constantly tweaking, optimizing, and perfecting – it's a never-ending quest for the perfect battery. Their patent fillings are very general, but one can get an idea of what is going on.

The Bottom Line: It's a Micron-Sized Miracle

Building SSB cells is a mind-blowing feat of engineering. It's a testament to how far we've come in our ability to manipulate matter at the atomic level. It's a combination of materials science wizardry, precision engineering that would make a watchmaker weep, and a healthy dose of "we can do this, even if it kills us" determination.

It's expensive, it's complex, and it's challenging as hell. But the potential reward? A revolution in energy storage that could power a cleaner, more sustainable future. And that, my friends, is worth fighting for – one meticulously welded, perfectly aligned, dust-free micron at a time.

Think of SSBs like a perfectly layered cake. Instead of fluffy sponge and creamy frosting, we're talking about incredibly thin layers of solid materials, all stacked precisely on top of each other. One of those layers is the solid electrolyte – the superhighway for ions (the tiny charged particles that make a battery...well, a battery). Unlike the liquid electrolytes in your current phone battery, which are a bit like a free-flowing river, this solid electrolyte is more like a solid, perfectly paved road.

The problem? Even the tiniest crack in that road, a microscopic pothole invisible to the naked eye, can cause a traffic jam of ions. And a traffic jam in a battery means reduced performance, overheating, and potentially, even a dangerous failure. We're not aiming for "mostly good" here; we need perfection.

Enter the "Guardians of Quality" – not a superhero team (though they are pretty heroic in the battery world), but a squad of super-smart diagnostic tools. These are the technologies that make sure our perfectly layered cake is, well, perfect.

Imagine having a superpower that lets you see inside things, not just on the surface. That's what these Guardians do for SSBs. One of the star players is a souped-up version of ultrasound – yeah, the same tech that lets doctors see babies in the womb. But instead of fuzzy images of tiny humans, we're getting ultra-high-resolution "sound pictures" of the battery's guts.

Think of it like this: High-frequency sound waves are pulsed into the battery. If everything's perfect, the waves bounce back in a predictable way. But if there's even a microscopic air bubble, a hairline crack, or a spot where the layers aren't perfectly bonded (a "delamination" – like when your cake layers start to separate), the sound waves scatter differently.

Now, here's where it gets really cool. Instead of relying on a human to squint at these sound-wave patterns (which, let's be honest, would look like abstract art to most of us), we bring in the AI brains. We're talking about machine learning – teaching a computer to become a master defect detective.

It's like showing a puppy thousands of pictures of cats and dogs until it can tell the difference between a chihuahua and a tabby with 100% accuracy. Except in this case, the AI is learning to differentiate between a perfect SSB and one with even the slightest flaw, based on those complex ultrasound patterns. And it's not just learning to spot problems; it's learning to characterize them. "That's a crack, it's 0.1 millimeters long, and it's located right...here."

This is a big deal, and not just in some theoretical, lab-coat kind of way. Big players like Samsung SDI are already using these AI-powered Guardians in their factories. It is letting these companies move past a simple test, and toward quality.

Why is this so crucial for SSBs? Because that solid electrolyte is unforgiving. Liquid electrolytes can kind of flow around minor imperfections. But a solid? A crack is a crack. And that crack can be a disaster waiting to happen.

But the AI isn't just a fault-finder; it's a teacher. It can actually predict potential problems before they even occur, and then can let engineers change parameters during manufacturing to minimize issues.

The AI can also provide information, so the engineer can go and look at the stress point, in which the issue is occurring.

Thanks to these Guardians of Quality, powered by the magic of AI, we're moving closer to a future where SSBs can safely and reliably power everything from our phones to our cars. It's about building batteries that aren't just powerful, but also incredibly robust and dependable – batteries we can truly trust.

From Grandma's Kitchen to Powering the Planet: The Battery Scaling Saga

Okay, picture this: You've cracked the code. You've got the ultimate battery tech – the holy grail that's going to make electric everything a reality. It works like a charm in the lab, a beautiful little prototype humming with potential. But... there's a tiny problem. You need to make, oh, about a billion more. And they all need to be exactly the same, cheap as chips, and ready to roll, yesterday.

Welcome to the wild, woolly world of battery scaling. It's less "Eureka!" and more "Oh, crap... how are we going to do this?" It's like going from perfecting Grandma's legendary cookie recipe (that one that uses a secret pinch of something special) to supplying every single bakery on the planet. Suddenly, that rickety old hand-cranked mixer and your trusty oven aren't going to cut it. You need an industrial revolution in your kitchen.

That's where the "modular" idea comes in. Think of it like LEGOs, but for battery factories. Instead of one gigantic, terrifyingly complex machine that does everything (and breaks down in a spectacularly expensive fashion), you build a bunch of smaller, specialized stations. One for mashing up the battery "dough," another for shaping it, another for "baking" it to perfection, and so on.

Tesla's Shanghai Gigafactory is like the rockstar example of this. They didn't just build a bigger oven; they built a whole city of coordinated ovens, each doing its part. And the brilliant thing is, if one oven goes on the fritz, the whole city doesn't grind to a halt. You can fix that one oven, or even add more ovens, without disrupting the cookie- (er, battery-) making flow. It is not magic, it is smart.

But let's not kid ourselves: building these battery mega-factories is expensive. We're talking Scrooge McDuck levels of money, swimming in vats of cash. It's not just the cost of lithium and cobalt (which, let's face it, aren't exactly pocket change). It's the everything else: the giant robots, the super-clean rooms, the army of PhDs making sure every battery is a twin of the last, and the sheer amount of power it takes to run the whole shebang. Every. Single. Penny. Counts. We're talking about shaving fractions of a cent off the cost of each battery – a game of microscopic margins that can make or break a company.

Northvolt is now taking a similar approach. They took note of lessons other companies are learning, and are now charting their own course.
It all comes down to constant improvement. Watching the data, and making the right changes, when they become needed.

The Shanghai Gigafactory wasn't just a carbon copy of Tesla's original factory in California. It was like taking Grandma's recipe and adapting it to use ingredients you can actually find in Shanghai, figuring out the local rules, and teaching a whole new crew of bakers how to make those cookies perfectly. They teamed up with Chinese suppliers, learned the local lingo (both literally and figuratively), and squeezed every ounce of efficiency out of the process.

Scaling up is never a simple "copy-paste" job. More like a "copy-paste-adapt-sweat-innovate-and-pray-it-all-works" kind of situation. It's a high-stakes game of engineering chess, where you need to be a master of materials science, a financial wizard, a supply chain guru, and a bit of a fortune teller, all rolled into one.

The companies that figure out how to build these battery Gigafactories – efficiently, affordably, and at lightning speed – are the ones who will literally power the future. It's a race, and the stakes are the planet. No pressure, right?

Economic Ripples of an Electric Shift

Electric Avenue: Ditch the Pistons, Grab the Future (and a Sweet New Job!)

Hold on to your hats, folks, because the robots aren't coming for your jobs – they're bringing new ones! Forget the oily rags and wrench-turning of yesterday's auto industry. We're plugging into a whole new world, a world buzzing with electric vehicles (EVs), and it's not just about saving polar bears (though that's pretty cool too). It's about a job boom so big, it's practically sparking! We're talking potentially 700,000 new jobs globally by 2030 – enough to fill a whole lotta football stadiums.

Imagine this: we're not just building cars anymore, we're crafting the future of transportation. Someone's gotta design those sleek, silent EVs that make you feel like you're driving a spaceship. Someone's gotta build the super-batteries that power them – think less AAA, more powerhouse. And someone's gotta make sure there's a charging station on every corner, so you're never stranded with a dead battery and a sad face. It's a whole new ballgame, a brand-new ecosystem, and it's desperate for skilled people.

Now, let's talk about the real heroes of this story: the workers. Take Michigan, the legendary home of American muscle cars. For generations, these folks have been the masters of the internal combustion engine – they could probably rebuild a carburetor blindfolded. But the times, they are a-changin'.

We're asking these incredible workers to learn a whole new language, a digital dialect of electric motors, battery wizardry, and power electronics. It's like going from writing with a quill pen to mastering a supercomputer. It's a challenge, no doubt, but it's also a golden ticket to a brighter future.

This is where the magic of retraining comes in. We're not tossing experience out the window; we're upgrading it! Think of it as a software update for your career. The International Labour Organization (ILO) – those smart folks who think about this stuff – are all about "just transitions." That's a fancy way of saying, "Let's make sure everyone gets a fair shot at this electric revolution." It means investing in training, offering support, and basically giving workers the keys to unlock these awesome new jobs.

And guess what? It's already happening! Look at Volkswagen in Chattanooga, Tennessee. They're not just churning out EVs; they're transforming their workforce. They're putting their money where their mouth is, giving their employees the skills to build the cars of tomorrow. That's not just good business; it's good karma. It's a win-win-win: for the workers, the community, and Mother Earth.

This isn't a simple job swap; it's a job explosion. And many of these new gigs come with fatter paychecks and a chance to be part of something truly groundbreaking. Think about all the pieces of the EV puzzle:

   Battery Bonanza: From digging up the raw materials to building those high-tech battery packs, this is a goldmine of opportunity.
   Charging Champions: We need a charging station network that would make Tesla proud – and that means jobs, jobs, jobs.
   Code Crunchers & Circuit Slingers: EVs are basically rolling computers, so we need brilliant programmers and electronics engineers to keep them humming.

Innovation Ignition: This technology is evolving faster than you can say "lithium-ion," so there's a constant need for bright minds to push the boundaries.

The Recycling Renaissance Reusing and repurposing those batteries is going to create a huge demand. The circular economy, creating the loop.

The EV job revolution isn't some pie-in-the-sky dream; it's happening right now. It's a chance to breathe new life into manufacturing towns, create fantastic careers, and build a future that's not just greener, but also brighter for everyone. It's an opportunity to learn, grow, and be part of something truly world-changing.

So, ditch the doom and gloom, friends. Grab your metaphorical (or literal) wrench, plug into the future, and get ready for the ride of your life. It's electric, and it's awesome!

The ICE Age is Over: A Gearhead's Goodbye (and Hello?)

For a century, the automotive world was a symphony of explosions. Not scary ones, mind you – controlled explosions, thousands of them per minute, all orchestrated by a beautiful, greasy, complex machine: the Internal Combustion Engine (ICE). We gearheads loved it. The rumble, the roar, the smell of gasoline on a crisp morning... it was our soundtrack. And for the companies that supplied the parts for this mechanical orchestra? They were the rock stars, the legends, the ones who kept the music playing.

But the times, they are a-changin'. Bob Dylan knew it, and the auto industry is living it. The electric vehicle (EV) revolution isn't just knocking on the door; it's kicked it down, walked in, and is currently raiding the fridge. And for those legacy suppliers, the folks who made their fortunes on pistons and fuel injectors? Well, it's a bit like being a blacksmith in the age of the automobile. Suddenly, your meticulously crafted horseshoes aren't in high demand.

This isn't a slow fade, either. It's more like Wile E. Coyote running off a cliff – that moment of suspended disbelief before the splat. We're talking about potential revenue drops that could make even the most seasoned CEO break out in a cold sweat. 30%? That's not a haircut; that's a full-body shave.

Imagine your favorite local bakery, the one that's been making the best sourdough bread for generations. Suddenly, everyone wants gluten-free, keto-friendly, air-puffed rice cakes. That bakery could adapt, learn new recipes, buy new ovens... or they could stick to their sourdough and slowly fade away. That's the dilemma facing many ICE parts suppliers.

Even the giants are feeling the heat. Take Bosch, for example. They're practically synonymous with the ICE – their name is practically etched onto every spark plug ever made. But even they are sprinting towards the EV future, throwing money at battery tech and electric motors like it's going out of style (because, for them, the old style is). It's like seeing your grandpa suddenly become a TikTok star – unexpected, but you gotta respect the hustle.

McKinsey, those sharp-dressed consultants with their fancy charts, are basically screaming from the rooftops: "Batteries! Batteries! Batteries!" The demand is exploding faster than a popcorn kernel in a microwave. It's a gold rush, a land grab, a feeding frenzy... pick your metaphor.

And while the ICE parts guys are facing a dust bowl, the battery folks are swimming in champagne.

This isn't just about business; it's about people. Think about the engineers who spent their careers perfecting the art of the fuel injector, the factory workers who can assemble an exhaust system blindfolded. What happens to them? The smart companies will retrain, retool, and find new roles for these skilled folks. The less-than-smart ones... well, let's just say there will be some sad stories.

The automotive world used to be a predictable highway, a straight shot down Route 66. Now? It's a chaotic off-road rally, full of unexpected twists, turns, and the occasional mud pit. The quiet hum of electric motors might be replacing the roar of the V8, but the underlying human drama – the struggle to adapt, innovate, and survive – is more intense than ever. It's a nail-biting race, and frankly, I can't wait to see who crosses the finish line first. And what kind of crazy vehicle they'll be driving.

Breathe Easy, Bank Better: Why Clean Air is the Ultimate Economic Stimulus

We've all heard the rallying cries: "Save the Planet! Go Green!" And those are essential. But let's be honest, sometimes saving the planet feels, well, big. Let's talk about something a little closer to home, something you feel in your lungs and in your bank account: clean air. It's not just about pretty sunsets and happy polar bears; it's about cold, hard cash. And we're not talking pocket change – we're talking a potential economic windfall.

Think of polluted air like a sneaky, silent tax. It's a tax on your health, a tax on your productivity, and a tax on our entire society. It makes us sick. Like, really sick. We're talking about the kind of sick that lands you in the hospital, keeps you home from work, and forces you to shell out for expensive medications. It's a drag on families, a burden on businesses, and a massive weight on our already-overwhelmed healthcare system.

Imagine $50 billion, by 2040 in the U.S. healthcare could be saved, just by making our air cleaner.

Now, flip the script. Imagine taking a deep breath of truly clean air. Not just "not-smoggy," but actually clean. That's not just good for your lungs; it's like hitting the economic jackpot. The folks at The Lancet Planetary Health – and these are serious scientists, not just tree-huggers (though they probably like trees!) – have crunched the numbers. They've shown that cleaning up our act, air-wise, isn't just feel-good; it's fiscally responsible.

Fewer kids gasping for air with asthma. Fewer adults clutching their chests with heart problems. Less strain on doctors, nurses, and hospitals that are already running on fumes. This isn't pie-in-the-sky dreaming; it's based on real data, real people, and real dollars and cents.

Take London, for example. They rolled out this thing called the ULEZ – the Ultra Low Emission Zone. Basically, they said, "If your car is a smoke-belching monster, you're going to pay extra to drive in the city center." Now, some folks grumbled (change is always hard!), but guess what? The air got cleaner. And, surprise, surprise, people started getting healthier.

Fewer sick days. Less pressure on the NHS (their national health service, which, let's face it, is constantly teetering on the brink). It's still early days, but the signs are clear: cleaner air equals a healthier bottom line. It saves money.

And London isn't alone. Cities and countries all over the world are waking up to this simple truth: Investing in clean air – through renewable energy, electric cars, and holding industries accountable – isn't just "environmentalism." It's smart economics.

So, let's ditch the idea that clean air is some expensive luxury. It's the opposite. It's an investment. A down payment on a healthier, wealthier future. The "clean air dividend" isn't just a catchy phrase; it's a powerful reality. It's about recognizing that the cost of inaction – the cost of continuing to breathe dirty air – is far, far greater than the cost of cleaning it up.

It's a win-win. A healthier population and a healthier economy. And that's something we can all, quite literally, breathe a sigh of relief about.

The Electric Dream Machine: Fueled by Billions

Forget chrome and horsepower. The real engine of the EV revolution? Cold, hard cash. And in 2023, that engine was roaring, fueled by a tidal wave of $15 billion in venture capital, according to the data wizards at PitchBook. Imagine that much money – it's not just stacks of bills; it's the lifeblood of ambition, flowing into the veins of companies like Arrival, who are daring to reimagine how we move things around.

Think of venture capitalists not as stuffy bankers in suits, but as modern-day gold prospectors, only instead of pickaxes and pans, they wield spreadsheets and PowerPoints. They're placing massive bets on the electric frontier, shouting, "We believe in this electric future, and we're putting our fortunes on the line!" It's the kind of money that turns a wild idea sketched on a napkin into a gleaming, whisper-quiet electric van zipping down your street.

This isn't some isolated phenomenon. It's a full-blown stampede. The old guard of the auto industry is scrambling to go electric, spurred on by both the carrot of opportunity and the stick of increasingly strict environmental rules. It's a "get on board or get left in the dust" kind of moment, and that urgency is like catnip for investors.

Fisker's story is the perfect example. Their IPO – basically, their grand debut on the Wall Street stage – wasn't just about their eye-catching Ocean SUV. It was a resounding "YES!" from the financial world, a collective nod of approval for the entire EV concept. It was like the cool kids finally inviting the electric car nerds to the party, and the nerds brought all the snacks (and the future of transportation).

But hold your horses (or, should we say, your electric steeds?). This isn't a fairy tale where everyone gets a trophy. Venture capital is a high-stakes poker game, and for every Fisker celebrating a win, there are countless other startups whose dreams will sadly turn to electric dust. That $15 billion represents a mountain of hope, but also the potential for a whole lot of heartbreak. The EV playground is getting crowded, and the competition is brutal. Not everyone will cross the finish line.

The real champions will be the ones who can turn that mountain of cash into something real. We're not just talking about flashy designs; we're talking about cars that are practical, reliable, and affordable for everyday people. It means building a charging network that makes "range anxiety" a thing of the past and convincing skeptical drivers that electric isn't just the future – it's right now.

This capital surge is the foundation. It's paying for the brainpower, the late-night coding sessions, the flashy marketing, and the sprawling factories. It's funding the birth of new technologies, new materials, and new ways of building things. It's a colossal effort, and the potential rewards – and consequences – are enormous.

The electric future isn't just some engineer's pipe dream; it's a shared vision, powered by the belief (and the billions) of investors who see a cleaner, quieter, and more exciting world on wheels. Its the investors placing a finical backing on the future of transport.

Greening the Grid: Sustainability in Focus

From Dust to Destiny: Your EV's Secret Life (and Why Solid-State Batteries Are Game-Changers)

Let's be honest, when you see a sleek EV gliding down the street, you're probably thinking, "Zero emissions! I'm practically saving the planet with every mile!" And you know what? You're partly right. But that silent, electric glide hides a much bigger story – a journey that starts long before you slide into the driver's seat and extends way past the day you trade it in. It's a story of rocks, rare metals, massive factories, and hopefully, a second life.

We're talking "cradle-to-grave," baby! And no, it's not about baby EVs (though that would be a cute concept). It's about understanding the entire environmental impact of your electric ride, from the moment the first chunk of lithium is ripped from the earth to the day its battery components (hopefully) get a new purpose. It is called a Life Cycle Assessment or LCA. Think of it as a full-body scan for your car's eco-footprint.

Now, here's where things get interesting. Right now, most EVs rely on lithium-ion batteries. They're the trusty workhorses, but they're also kind of like that friend who's awesome but has a few hidden baggage fees. Mining all those materials – lithium, cobalt, nickel – it's messy. It takes a lot of energy. And building those batteries? Another energy-guzzling process.

Enter the hero of our story: Solid-State Batteries (SSBs). Imagine a battery that's like a super-efficient, long-lasting, incredibly safe superhero. That's the promise of SSBs.

Scientists, the real-life Tony Starks of the battery world, are geeking out over this. There was even a super-important paper in Environmental Science & Technology (it's basically the Rolling Stone magazine for environmental nerds) that said EVs with SSBs could have up to 40% lower lifecycle emissions than the ones with the current batteries. 40%! That's like going from a gas-guzzling Hummer to, well, a much more eco-friendly EV.

How do they pull off this magic trick?

    More Power, Less Stuff: SSBs are energy-dense. Think of them as tiny, incredibly efficient powerhouses. You need a smaller battery to go the same distance, which means less mining, less processing, less everything.
    The "Secret Sauce" is Gone: Traditional batteries have a liquid electrolyte, which requires some pretty energy-intensive manufacturing steps. SSBs use a solid electrolyte, skipping some of those messy processes.
    Built to Last: SSBs are projected to live longer and be safer. That means fewer replacements and a smaller long-term environmental impact.

Think about Volvo and their C40 Recharge. They're already super open about their car's environmental footprint (which is refreshing, right?). They track everything. Imagine that car, but with an SSB heart. The numbers would be even more impressive.

Now, let's be real. Even SSBs aren't perfect. Mining is still mining, even if we need less of it. We absolutely need to make sure it's done responsibly – fair wages, protecting the environment, the whole nine yards. And we need to get way better at recycling those batteries at the end of their life. The good news is, there's a ton of research happening on that front, with scientists coming up with ingenious ways to reclaim those valuable materials.

The future of EVs isn't just about plugging in. It's about the whole story, from that first speck of dust to that final, hopeful rebirth. SSBs are a massive step towards a cleaner, more sustainable future, making that story a whole lot more inspiring. It's a continuous evolution, a constant striving for better. And that's something worth getting excited about.
It is not perfect, but that 40% is a big deal.

>From E-Waste Graveyards to Battery Nirvana: The Solid-State Resurrection

Let's be honest, the phrase "solid-state battery" sounds like something ripped from a sci-fi flick. But these futuristic powerhouses are about to be everywhere – powering our phones, laptops, and yes, even those sleek electric cars zipping down the highway. They promise a world of faster charging, longer-lasting power, and a safer battery experience. But... what happens when these super-batteries finally kick the bucket? Do they join the depressing mountains of e-waste, a monument to our throwaway culture?

Thankfully, the answer is a resounding, hopeful no. We're not just talking about responsible disposal; we're talking about a full-blown battery resurrection.

Imagine a phoenix, rising from the ashes, reborn and ready to soar. That's the future of solid-state batteries (SSBs), thanks to some seriously ingenious science and a growing commitment to a "circular economy" – basically, making sure nothing goes to waste.

Forget harsh chemicals and energy-guzzling furnaces. The real magic happens at the microscopic level. Picture this: tiny, incredibly precise "molecular scissors" – enzymes, actually – delicately snipping apart the complex innards of a spent SSB. These aren't clumsy hacks; they're like a microscopic surgical team, expertly separating the valuable bits: lithium (the energetic heart of the battery!), cobalt, nickel, and other precious materials.

Companies like Redwood Materials are leading this charge, and their pilot programs are blowing minds. They're recovering a whopping 95% of the materials from these dead batteries using these enzyme-powered processes. That's not just impressive; it's revolutionary.

Why does this matter so much? Because mining these materials in the first place is often a messy, environmentally damaging, and sometimes even ethically questionable business. By creating a "closed loop" – where old batteries become the raw materials for new ones – we're not just being eco-friendly; we're building a more secure and independent supply chain. Think less reliance on faraway mines and more resourcefulness right here at home.

And this isn't some pie-in-the-sky academic theory. Green Chemistry, the journal that's all about sustainable solutions, is buzzing with research on these enzymatic recycling methods. And major players like BMW are already putting their money where their mouth is. BMW is building a closed-loop system for their battery materials, aiming to feed those recovered goodies straight back into making brand-new batteries. This isn't a "someday" dream; it's happening now.

So, what does this mean for you, me, and our gadget-obsessed world? It means that the electric future we're charging towards (pun intended!) can be surprisingly sustainable. It means that the devices we love, and the electric vehicles we'll soon be driving, won't become another environmental burden. Their components will get a second chance, a third chance, maybe even a fourth!

The idea of a truly circular battery economy – where nothing is wasted and everything is reused – is no longer a fantasy. It is within reach. It's a future where technology and nature work together, not against each other. Where old batteries don't die; they're reborn. And that's a future worth plugging into.

The Green Dream's Dirty Secret: Can We Mine Our Way to a Better Future – Responsibly?

We're all buzzing about the electric car revolution, the solar panel boom, the promise of a greener tomorrow powered by technology. It's exciting! We picture sleek Teslas gliding silently down the road, wind turbines gracefully turning against a clear blue sky. But hold on a second. There's a hidden chapter to this story, a gritty reality that's far removed from the glossy brochures and marketing campaigns.

The truth is, the guts of these amazing technologies – the batteries, the microchips, the solar cells – rely on raw materials like lithium and cobalt. And those materials? They often come from places where "sustainable" feels like a cruel joke. We're chasing a clean future, but the starting line is often incredibly messy.

Imagine this: You're in the Atacama Desert in Chile, one of the driest places on Earth. The sun beats down mercilessly. Here, beneath the cracked earth, lies "white gold" – lithium, the lifeblood of electric vehicle batteries. The traditional way to get it? Pump up vast amounts of salty, ancient water from underground reservoirs, and let the sun do the work of evaporation. It's like leaving a giant, salty footprint on a landscape that's already gasping for water.

Local communities, who depend on that precious water, are left struggling. The very thing that's supposed to save the planet is, in a way, stealing from it. It's a heartbreaking paradox.

But there's a glimmer of hope. Smart people are working on ways to do things differently. Think of it like this: instead of draining the whole bathtub to get the rubber ducky, we're trying to find a way to just scoop out the ducky. That's the idea behind Direct Lithium Extraction (DLE). It's about being surgical, precise, taking only what we need and leaving the rest. Even Tesla, the poster child of the EV world, is looking at ways to get lithium from clay in Nevada, trying to minimize their impact (they've said as much, and it's all over the industry news – they want to be good guys, at least in this respect!). It's not perfect yet, but it's a step in the right direction.

Now, let's jump to the Democratic Republic of Congo. This country holds a treasure trove of cobalt, another key ingredient in those powerful batteries. But the picture here is even more troubling. Much of the cobalt is mined by hand, in conditions that are often dangerous, and, tragically, sometimes involve children. Imagine that – the device in your pocket, the car in your driveway, potentially linked to such a heartbreaking reality.

This is where technology can play a truly amazing role. Imagine a digital fingerprint for every ounce of cobalt, tracking its journey from the moment it leaves the earth to the moment it powers your phone. That's the promise of blockchain. It's like a super-detailed, unbreakable chain of custody.

Think of it as a spotlight shining on the entire supply chain. It can't magically solve all the problems, of course. We still need people on the ground, making sure things are done right. But blockchain can give us a level of transparency we've never had before. It lets us see the story behind the materials, and demand better.

Ultimately, this isn't just about fancy tech or clever solutions. It's about a fundamental shift in how we think. It's about recognizing that "cheap" often comes at a very high price – a price paid by the environment, by communities, by real people.

"Mining with conscience" is about understanding that true sustainability is an investment. It's about choosing the right way, not just the easiest or cheapest way. It's about making sure that the green revolution doesn't leave a trail of destruction in its wake. It's about having our cake, and eating it too, in a way where the earth does not take the loss.
It is a long road, but it is the most important.

Japan's Battery Game: It's Like 'Mario Kart' for Eco-Efficiency

Imagine a real-life video game, but instead of racing karts, companies are racing to build the greenest battery. That's the vibe of Japan's Top Runner Program. It's not some dusty, boring government regulation; it's a high-stakes competition where the planet wins.

Forget minimum standards – that's like playing the game on "easy" mode. Top Runner is like setting the difficulty to "expert." Japan looks at all the batteries out there, finds the absolute best, most eco-friendly one (the 'Toad" of batteries, if you will – surprisingly fast and efficient!), and says, "Okay, everyone, this is the new standard. Beat it!"

It is a constant level up.

This isn't a suggestion; it's the law. And it's got teeth. Companies like Panasonic aren't just casually strolling towards sustainability; they're sprinting. They have to, or they'll get left in the dust. It's like that feeling when you see a blue shell coming in Mario Kart – you gotta boost!

The journal Energy Policy is like the game's official strategy guide, full of data showing how this program is actually working. It is pushing and speeding up the process.

Why all this fuss about batteries? Because they're the unsung heroes (and sometimes villains) of our modern world. They power everything, but making them can be a dirty business – like mining for rare resources is a bit like digging up the Mushroom Kingdom, not always pretty.

Top Runner forces companies to think about the whole battery life cycle. It's not just about how long it lasts in your phone; it's about where the materials came from, how much energy it took to make it, and what happens when it's finally dead. It's like asking, "Did Mario really need to stomp all those Goombas?" (Okay, maybe that's a stretch, but you get the idea.)

Now, it's not a perfect system. Smaller companies might feel like they're stuck in the slow lane, struggling to keep up with the big players. And there's always the risk of companies trying to "cheat" the system, focusing on the rules instead of the spirit of the game. The system needs the players' commitment.
But – and this is a big but – Top Runner is a bold move. It's Japan saying, "We can be eco-friendly and economically competitive." It's a challenge to the world, a signal that we can build a better future, one supercharged, eco-friendly battery at a time. And, just like in any good game, there's always a new level, a new challenge, a new "Top Runner" to chase. The challenge will be never over.

Tomorrow's Blueprint: Visionary Projects

EVs of the Future: From Sci-Fi Sketches to Your Garage – Will the Afeela Actually Work?

We've all been there. Drooling over concept EVs at auto shows. They look like they've been beamed down from a more stylish, technologically advanced planet. All curves and impossible angles, with interiors that make your current car feel like a horse-drawn buggy. They promise a future that's always just around the corner, perpetually five years away. But how do these beauties actually become, you know, real cars?

Enter the Afeela, the electric lovechild of Sony and Honda. It's not just another tease. This thing is packing some genuinely wild tech that might, might actually end up in a car you can buy. And that's where things get interesting.

Let's be honest, concept cars are usually about as practical as a chocolate teapot. They've got ground clearance that would make a speed bump nervous, doors that'd require a crane to open in your average parking garage, and wheels borrowed from a monster truck. The Afeela, bless its heart, seems to be making an effort to be, well, usable. The designers have clearly been to an actual, real-world parking lot. Bravo!

But the real party is happening inside. Holographic displays. I'm not kidding. The Afeela's prototype is throwing information into thin air. Think navigation arrows that float magically on the road ahead, or entertainment that just appears for your passenger (no more fighting over the aux cord!). It's the kind of stuff that used to belong exclusively to Captain Kirk and the Enterprise.

Now, hold your horses (or, I guess, your electric unicorns). I've been to enough automotive engineering shindigs (the Automotive Engineering Expo is a good one, lots of smart people in lab coats) to know that holographic displays in cars are still basically toddlers in the tech world. They've got some growing up to do. Think about it: can you actually see the hologram when the sun's blasting through your windshield? Will it suck your battery dry faster than a teenager on TikTok? And, most importantly, will it distract you so much you end up driving into a lamppost? Making this tech cool and safe is a challenge that would make even Elon Musk sweat a little. The concept is very revolutionary, no doubt about it.

And then there's the engine – or, rather, the lack of a traditional one. The Afeela whispers sweet nothings about solid-state batteries (SSBs). These are the legendary Excalibur of EV batteries. Compared to the lithium-ion batteries we've got now, SSBs are supposed to offer way more range, charge faster than you can say "supercharger," and be less likely to spontaneously combust (which is always a plus).

But, again, let's tap the brakes a bit. The same folks at those engineering conferences make it clear: SSBs are amazing on paper, but making enough of them to power the world's cars? That's a whole different ballgame. We're talking about refining manufacturing processes, bringing down costs (because who wants a car that costs more than a house?), and making sure these batteries last longer than your last smartphone.

So, the Afeela is kind of a fascinating paradox. It's a glimpse of a genuinely exciting future, but it's also a reminder that getting there is a messy, complicated, step-by-step process. It's like watching a really promising movie trailer – you're hyped, but you know the actual movie might have some plot holes. It beautifully straddles that line of the fantastic with realism.

SSB Leap Forward: Halide Electrolytes – Forget the Tortoise, We're Talking Cheetah-Speed Charging!

Solid-state batteries (SSBs) – they're the mythical unicorn of the battery world. Everyone wants one: longer-lasting EV charges that'll get you from LA to Vegas without a sweat, phone batteries that laugh in the face of a full day's use, and, blessedly, no more fiery laptop surprises. But the journey to SSB-land has been less "majestic gallop" and more "stumbling through quicksand." The biggest culprit? The electrolyte – the crucial middleman in the battery's energy game.

Think of the electrolyte as the battery's superhighway for ions. In your current lithium-ion battery, that highway is a liquid – it works, sure, but it's also the reason your phone occasionally gets hotter than a jalapeño and why "battery fire" is a phrase we even have to know. SSBs ditch the liquid for a solid, which is inherently safer. The problem? Finding a solid that's also a speed demon for ion transport. That's been the holy grail search.

But hold onto your hats, folks, because we might have just found Excalibur. Recent breakthroughs with halide electrolytes are causing serious buzz, and for good reason. These aren't your grandma's electrolytes (unless your grandma is a cutting-edge materials scientist). We're talking about materials packed with elements like chlorine, bromine, and iodine, and they're showing off some seriously impressive skills in the lab.

The headline? Double the charging speed. Let that sink in. Imagine plugging in your EV and it's ready to roll in half the usual time. Your phone? Fully juiced in the time it takes to make a cup of coffee. This isn't just a slight improvement; it's a potential game-changer.

The brainiacs over at Science Advances have been publishing the juicy details, revealing why these halide electrolytes are such overachievers. It's not just about raw speed; it's about stability too. These halides are like the perfect party guests – they get along really well with the battery's electrodes, creating a strong, stable connection. This means less battery degradation and a longer lifespan. Because a battery that charges fast but dies young is about as useful as a chocolate teapot.

And this isn't just some theoretical, "in-a-lab-far-far-away" kind of thing. Real-world giants are taking notice. Toyota, a company that usually moves with the speed of a cautious glacier when it comes to new tech, is making bold promises about SSBs for 2025. You can bet your bottom dollar that halide electrolytes are playing a starring role in their plans.

Of course, it's not a halide-electrolyte-only party. There are other solid electrolyte contenders out there – sulfides, oxides, the whole gang. But the halide breakthroughs, especially those being rigorously tested at the prestigious Argonne National Laboratory, are sending shockwaves of excitement. Argonne isn't just taking pretty pictures; they're deep-diving into the how and why, dissecting these materials under every imaginable condition.

So, should you start holding your breath for a halide-powered phone next week? Probably not. But should you be excited? Absolutely! This is more than just a baby step; it's a sprint forward. We're talking about a potential paradigm shift in battery technology, moving us closer to a world where our devices are safer, more powerful, and finally keep up with our increasingly electrified lives.

And keep your ears to the ground. More juice about the resarch, is undoutably on its way.

Sun-Kissed Rides: When Your EV Sings the Same Song as Your Solar Panels

Let's be honest, that little voice in the back of your head always whispers when you plug in your EV. "Is this really clean energy? Or am I just burning coal in a slightly more fashionable way?" It's like ordering a salad and then realizing it's drenched in creamy, calorie-laden dressing. The good intention is there, but the reality... eh.

But what if your electric car could actually sip sunshine? What if your home battery wasn't just a backup, but a tiny, sun-powered orchestra conductor, perfectly harmonizing with your car's energy needs? We're not talking about some sci-fi fantasy; we're talking about the dawn of Power Harmony – where EVs and renewable energy aren't just dating, they're married and making beautiful, sustainable music together.

Picture this: You're cruising in a Lightyear One. It's not just an EV; it's a four-wheeled sunflower, constantly turning its face to the sky. Five square meters of solar cells, cleverly disguised as a sleek, aerodynamic body, are drinking sunlight. They're not just providing a little "range extender" – they're promising up to 70 kilometers (43 miles) of free, sun-powered driving per day if the sun gods are smiling. Okay, okay, maybe you won't get that in Seattle in November, but even a cloudy day delivers some juice. It's like having a tiny, personal power plant – one that happens to look incredibly cool.

Now, let's zoom in on the Netherlands. These folks aren't just known for windmills and tulips; they're pioneers in energy innovation. Imagine this: Homes are transforming into mini power hubs, thanks to Solid State Batteries (SSBs) – think of them as super-efficient, high-tech energy piggy banks. Your solar-powered car rolls up, overflowing with sunshine energy. Instead of that energy disappearing into the ether, it flows into your home's SSB. Then, when the moon replaces the sun, your car can draw from that stored solar goodness. It's a closed loop, a beautiful energy dance, like a perfectly choreographed tango between your car, your house, and the sun. You've essentially told the traditional, potentially coal-fired power grid, "Thanks, but no thanks. I've got this."

This isn't just some engineer's fever dream fueled by too much coffee. It's backed by real science, the kind you find in those brainy renewable energy journals. Scientists are basically becoming energy matchmakers, figuring out how to make solar panels, EV batteries, and home storage systems communicate like old friends, sharing energy seamlessly and efficiently. Think of it as a smart grid on a micro-scale, orchestrated by some seriously clever algorithms.

The real magic of Power Harmony? It's about empowerment. We're not just passive consumers of energy anymore; we're becoming producers. Our homes, our cars – they're mini power stations, feeding energy back into the system. It's a revolution, a shift from massive, centralized power plants to a distributed network of solar-powered awesomeness. Less strain on the grid, less energy wasted in transmission, and a giant leap towards a smaller carbon footprint.

Now, let's address the elephant in the room: the price tag. Yes, solar-equipped EVs and cutting-edge batteries like

SSBs are currently a bit like that designer handbag – beautiful, desirable, but definitely an investment. And the technology is still learning to walk – it's not quite running marathons yet. Getting that perfect balance between sunshine collection, energy storage, and your car's thirst for power requires a delicate dance of technology.

But...(and it is a Big BUT), the wheels are turning (pun intended!). The Dutch trials, along with other pioneering projects, are showing us what's possible. We're moving beyond the era of "plug and pray" – praying that the electricity powering our EVs is actually clean. We're entering an age where our cars are active players in a truly sustainable energy ecosystem, a symphony conducted by the sun. And that, my friends, is a future worth getting charged up about.

Okay, buckle up, buttercup, because we're about to take a joyride into the future of electric vehicle charging, and it's way cooler than you probably imagine. Forget that mental image of plugging your car in like it's a giant toaster. We're talking about a full-blown, ground-up revolution, a complete reimagining of how we juice up our rides.

First stop: Megawatt Charging. Let's ditch the technical jargon for a sec. Current "fast" chargers? Think of them as a garden hose filling a swimming pool. Megawatt charging? That's the Niagara Falls of electricity. We're talking about ONE MILLION WATTS of raw power slamming into your vehicle's battery. Instead of waiting around for hours, you'll be back on the road with hundreds of miles of range after a quick coffee break. Seriously, it's that fast.

This isn't just about convenience for your daily commute. This is a game-changer for the big boys – semi-trucks, buses, the workhorses of our economy. Those long, grueling hauls that seemed impossible for electric vehicles? Megawatt charging makes them a reality. Think about the folks at places like the IEEE Power Electronics Society – they're the mad scientists in lab coats, the ones pushing the absolute limits of what's possible with electricity. They're the reason this isn't just a pipe dream.

But hold on, it gets even wilder. Imagine never plugging in at all. Sounds like something out of a sci-fi flick, right? Nope. We're talking about inductive roads, also known as dynamic charging. Picture this: the road itself is your charger.

Think of it like this: the road is embedded with special coils, like hidden treasure beneath the asphalt. As your car (which needs to be equipped for this, of course) glides over these sections, it's like a wireless phone charger, but on a massive scale. Your car sips energy while you drive.

Sweden, those eco-pioneers, are already making this a reality. And companies like Electreon? They're not just tinkering in labs; they're laying down real, honest-to-goodness electric roads where cars can charge on the fly. It's mind-blowing.

Imagine the freedom! Range anxiety? Poof! Gone. Charging stations? Potentially a thing of the past (at least on major routes). City buses could run 24/7, never needing to duck back to the depot. Delivery trucks could crisscross the country without stopping. And your own EV? It's like having an endless supply of fuel, magically appearing as you cruise along.

Plus. These electric roads have added safety. No more tripping hazards from stray charging cables.

Now, let's be real. This isn't going to happen overnight. Building this kind of infrastructure is a huge undertaking. We need to figure out how to seamlessly weave this technology into our existing roads. And we need everyone – car manufacturers, road builders, governments – to agree on standards, so everything works together smoothly.

But think about the payoff! A future where electric vehicles aren't just a "greener" option, but the obvious choice. A future where our roads themselves are part of a sustainable energy ecosystem. It's a future powered by pure ingenuity, a commitment to a cleaner planet, and, frankly, a healthy dose of "wouldn't it be cool if...?" thinking. The charging revolution is here, and it's happening at warp speed, both above ground and right beneath our tires. Get ready for the ride!

The Consumer Pulse: Adoption Dynamics

The Electric Slide: It's More Than Just Plugging In

Let's face it: We've all been there. You're cruising down the highway, and suddenly, a sleek, silent spaceship glides past. It's an EV, all futuristic curves and smug self-satisfaction. And for a fleeting moment, you feel a pang. Not just envy, but a whisper of... possibility. The EV revolution isn't about ditching gas; it's about ditching baggage – the rumble of the engine, the guilt at the pump, the feeling you're driving a dinosaur.

It's like this: remember that kid in school who always had the latest gadgets? The one with the laser pointer and the transparent Game Boy? EVs tap into that same primal desire to be ahead of the curve. Deloitte's research confirms it: we're not just buying cars; we're buying into an identity. We're saying, "I'm part of the future, and I look good getting there."

The Journal of Consumer Research calls this "identity signaling." Basically, your car is your billboard. And an EV, especially something like a Kia EV6 (which, let's be honest, looks like it escaped from a sci-fi movie set), screams, "I'm stylish, I'm eco-conscious, and I probably have excellent taste in coffee." It's not just transport; it's a vibe. It is a statement.

But (and there's always a "but," isn't there?), there's a gremlin lurking in the battery pack. It's the little voice that whispers, "What if you run out of juice on that road trip to Grandma's?" It's range anxiety, the EV equivalent of stage fright.

Imagine you're dating someone amazing. They're smart, funny, gorgeous... but they have a really weird quirk, like they only eat food that's purple. You'd be intrigued, maybe even excited, but also... a little nervous, right? That's range anxiety.

Deloitte's surveys keep hitting this same nerve. We're obsessed with charging. Where are the chargers? How long will it take? Will I be stuck in a desolate parking lot, battling tumbleweeds and boredom while my car sips electrons? These fears aren't always logical – the charging network is growing faster than a teenager's appetite – but they're real. They're rooted in a century of gas-guzzling habits.

It's a battle between our logical brain (which knows EVs are the future) and our lizard brain (which just wants to get home without drama). The head says, "Lower emissions, cheaper to run, cutting-edge tech!" The heart whimpers, "But... what if...?"

The EV market is at this thrilling, slightly awkward crossroads. It's like a first date – full of potential, but also a bit nerve-wracking. The "cool factor" is off the charts. The cars look amazing, and the idea of driving the future is intoxicating. The curb appeal is at an all-time high.

But conquering those charging jitters? That's the key to unlocking the whole shebang. It's not just about installing more plugs (though, please, do that). It's about empathy. It's about understanding that switching to an EV is like learning a new dance. It takes practice, a bit of trust, and maybe a reassuring instructor (or a really good user manual).

The companies that "get" this – the ones that can make us feel both excited and secure – will be the ones leading the electric parade. It's not just about wallets; it's about winning hearts and minds, one (smooth, silent) ride at a time.

The EV Revolution: Not One Size Fits All (Especially Wallets)

Forget the idea of a single, massive wave of EV adoption. It's more like a bustling harbor, filled with all sorts of vessels, each with its own captain and destination. To navigate this market, you need to understand the people at the helm, not just the horsepower under the hood. We're talking about a world of markets within markets.

The Fleet Commander: CFO in Disguise

Imagine Frank. Frank runs a plumbing business, "Frank's Flow Fixers." He's got a fleet of vans, and they're his workhorses, not his show ponies. Frank doesn't care about 0-60 times; he cares about 0-headaches-per-month. He is being pushed, though, by the city counsel. His choice between a gas van and an EV van is no different than the selection of the plumbing parts he picks.

For Frank, going electric is a spreadsheet symphony. It's all about TCO – Total Cost of Owning (and not breaking down). Fuel savings? Music to his ears. Reduced maintenance? A chorus of angels. Government throwing in some tax breaks? That's the encore.

Frank's not buying a dream; he's buying a tool. He's grilling the salesperson about charging logistics: "Can my guys plug these in overnight at the depot? What happens if one conks out on Route 5?" He's pouring over reports from the Market Research Society – they're basically his bible right now, showing him that other Franks are making the switch and not going bankrupt. He's learning from their early screw-ups – the charging station nightmares, the drivers who initially freaked out about running out of juice. Frank's a practical guy; he wants solutions, not slogans.

The Luxury Cruiser: Status Seeker with a Green Tint

Now, picture Isabella. Isabella doesn't need a car; she desires a statement. Her ride of choice? Maybe a Rolls-Royce Spectre, a whisper-quiet chariot of electric elegance, or perhaps a Porsche Taycan that pins you to the seat with silent fury.

Isabella isn't crunching numbers on fuel costs. She's basking in the glow of instant acceleration, the kind that makes your stomach do a little flip. She's reveling in the hushed cabin, a sanctuary from the city's roar. She's admiring the sculpted lines, the futuristic dashboard that looks like it was ripped from a spaceship.

For Isabella, this isn't about saving pennies; it's about signaling success, sophistication, and a certain eco-consciousness (even if her private jet usage might negate that a bit). It's an emotional purchase, fueled by the desire to be ahead of the curve, to be part of something exclusive. While Frank is fretting about kilowatt-hours, Isabella is choosing between custom leather interiors and ambient lighting colors. Her question isn't "Will it pay for itself?", but "Does it make me feel extraordinary?"

The Big Picture: A Mosaic, Not a Monolith

Frank and Isabella are just two characters in a much larger drama. The EV market isn't one big, homogenous blob. It's a vibrant mosaic, made up of countless individual needs, desires, and budgets. Whether it will be the government purchasing a military EV, or a family. Understanding these diverse "markets within markets" – from the pragmatic fleet manager to the aspirational luxury buyer, and everyone in between – is the key to unlocking the true potential of the electric vehicle revolution. It is about putting the right car with the proper consumer.

Okay, buckle up, because the EV revolution is here, and it's not some distant, futuristic fantasy. We're talking about "Sales Trajectories: Numbers Tell the Tale," and this tale is a page-turner, a real thriller, even!

We're staring down the barrel of 20 million electric vehicles sold by 2027. Let that sink in. Twenty. Million. That's not just a blip on the radar; that's a seismic shift, a tectonic plate rearranging the entire automotive world. We're witnessing the birth of a new era, a move from those quirky early adopters to your neighbor, your grandma, everyone potentially plugging in. This isn't just about cleaner air (though that's a huge win); it's about reshaping how we think about transportation, infrastructure, and even our daily routines.

And who's leading this electric charge? It's China, and they're not just leading, they're dominating. Imagine a race where one runner is so far ahead, they're practically lapping the competition. That's China, projected to snag a mind-blowing 60% of that 20 million EV market. Sixty percent! They're not just playing the game; they're basically writing the new rulebook, fueled by a potent cocktail of government support, a rapidly expanding charging network, and a population that's saying "yes!" to electric dreams. There seems to be some genuine enthusiasm, as well.

Now, I know what you're thinking: "Big numbers, fancy projections...show me the receipts!" We're not just spinning yarns here. This isn't some back-of-the-napkin calculation. This forecast is anchored in rock-solid research, primarily from the International Energy Agency's (IEA) Global EV Outlook. Think of the IEA as the Sherlock Holmes of the energy world – they're the go-to detectives for unraveling global energy trends. Their EV Outlook is the gold standard, the bible, for anyone wanting to understand where the electric vehicle market is headed. They're meticulously collecting data, poring over policy documents, tracking the latest tech breakthroughs, and building sophisticated models to predict the future. It's like having a crystal ball, but backed by hard science.

But we're not just staying in the realm of global forecasts. We're zooming in for a closer look, a real-world case study, and that's where Norway comes into the picture. Norway is like the EV world's rockstar, the early adopter who went all-in and is now reaping the rewards. They're the living proof that widespread EV adoption isn't just a pipe dream; it's happening now. They're light years ahead of the pack, with EVs making up a massive chunk of their new car sales.

Think of Norway as a living laboratory, a real-time experiment in electric mobility. By studying their journey, we can learn so much: How do incentives really impact people's buying decisions? When do people actually plug in their cars, and is the power grid holding up? What bumps in the road did they hit, and how did they smooth them out? Norway's sales figures, charging station data, even those little customer satisfaction surveys – they're all like pieces of a puzzle, helping us understand what a truly electrified future looks like.

So, bottom line? The numbers are shouting, not whispering. The EV revolution is on a rocket ship to the stars, with China holding the launch codes. And while the IEA's projections give us the grand, sweeping view, Norway's on-the-ground experience provides the close-up, gritty details. It's this powerful combo of big-picture analysis and real-world stories that paints such a vivid, compelling picture of where we're headed. It's a future that's zooming towards us faster than we ever imagined. And yes, it's absolutely going to be a fascinating ride to watch unfold!

Okay, let's ditch the textbook talk and get real about why you're seeing so many Tesla Model Ys zipping around – and how the government played a sneaky-smart role in that.

Imagine this: You're scrolling through Tesla's website, drooling over the Model Y. It's sleek, it's fast, it's electric. But then you see the price tag, and your wallet starts doing that nervous twitch thing. Oof.

Enter Uncle Sam, stage right, with a wink and a fat stack of cash (well, potential cash). He whispers, "Hey, buy that electric beauty, and I'll knock $7,500 off your tax bill." Suddenly, that Model Y isn't just a dream car; it's a financially savvy dream car.

That's the power of the $7,500 federal EV tax credit, folks. It's not just some boring policy detail; it's a straight-up game-changer. It's like the government saying, "We really want you to ditch gas guzzlers, and we're willing to put our money where our mouth is."

Why is the Model Y such a big deal in this story? Because it's the Goldilocks of Teslas. It's not the super-expensive Model S or X, but it's not the (still somewhat pricey) Model 3 either. It's just right for a lot of people – families who need space, commuters who want to save on gas, anyone who wants a taste of that Tesla magic without completely emptying their bank account.

And that $7,500? It's like a magical price eraser. It doesn't make the car cheap, but it makes it approachable. It turns "maybe someday" into "hmm, maybe now."

This isn't rocket science; it's basic human psychology. We like getting deals. We like feeling like we're making smart choices. And we really like getting money back from the government.

Think about what France does. They've got this "bonus-malus" thing – it's like a reward and punishment system for cars. Buy a clean car? Get a bonus! Buy a gas-guzzling monster? Get slapped with a penalty. It's like they're saying, "Be good to the planet, and we'll be good to your wallet." And guess what? It works. People are actually buying more eco-friendly cars because of it.

The cool thing is, this isn't just about individual buyers. It's like a domino effect. When lots of people start buying Model Ys because of the tax credit, Tesla (and other car companies) notice. They think, "Whoa, people really want electric cars!" So they start pouring more money into making even better and cheaper EVs. It's a win-win-win: We get cooler cars, the planet gets a break, and the government gets to pat itself on the back for a job well done.

Now, I'm not saying this is a perfect system. There's always fine print, and people love to argue about whether these tax credits are fair to everyone (do they mostly help rich people?) and if they'll last forever. The rules change more ofen then the weather!
But let's be honest, for now, it works. It's nudging us, one electric car at a time, towards a greener future. And that, my friends, is pretty darn cool. It proves that sometimes, all it takes is a little financial sweetener to make a big difference in the world.

Fortifying SSBs: Safety as Priority

The Fire Within (and How We Tamed It): The Cool Story of Solid-State Batteries

Forget ticking time bombs. We all know the old battery narrative: energy packed tight, a whisper away from a fiery meltdown. It's the tech equivalent of holding a dragon's breath in your pocket. But solid-state batteries? They're writing a new chapter, and the title is all about chill.

Think of it like this: your phone, your electric car, even your future electric airplane – they're all powered by a tiny, incredibly powerful engine. And like any high-performance engine, it needs to breathe. It needs a way to handle the heat, or it'll go from zero to inferno faster than you can say "thermal runaway." That's the scary phrase, the boogeyman of battery tech. It's a chain reaction, like dominoes of fire, and it's what keeps engineers up at night.

But here's where the heroes of our story come in: phase-change materials. Don't let the science-y name intimidate you. Imagine them as tiny, microscopic superheroes – thermal sponges. They're like the ice packs of the battery world, but way, way more sophisticated.

These aren't your average ice packs, though. These materials have a secret superpower. They can absorb a staggering amount of heat without even breaking a sweat (or, well, melting significantly). They do it through a bit of molecular magic. At a specific temperature, they change their state – think solid to liquid, or liquid to gas. And that transformation? It's like a black hole for heat. It sucks the energy in, keeping the battery cool and calm, even when things get intense.

The information provided states that LG Chem uses this technology, and it's cutting fire risks by a whopping 80%. That's not just a number; it's a revolution. It's the difference between a battery that's always on the verge of a meltdown and one that you can trust with your life, literally. It is a game-changer!

Imagine a car crash. Your electric vehicle is crumpled, the battery pack potentially damaged. In the old days, that was a recipe for disaster. But with these phase-change materials woven into the very fabric of the battery, it's a different story. They're the first line of defense, buying precious time, absorbing the shock of heat, preventing that fiery chain reaction from ever starting.

This isn't just lab coat theory, either. The Journal of Energy Storage – the place where the real battery nerds hang out – is buzzing about this. And LG Chem, a major player in the battery game, is already putting it into practice. That's like seeing a superhero movie become reality.

But the real artistry isn't just in the materials themselves; it's in how they're used. Imagine a microscopic web, a network of these heat-absorbing heroes, spread throughout the battery. They're not just slapped on; they're integrated, woven into the very DNA of the battery.

And it gets even cooler (pun intended!). Think smart cooling. Tiny sensors, like microscopic thermometers, are constantly on patrol. If a hotspot starts to flare up, they sound the alarm. Maybe they trigger a tiny, localized cooling response, like a miniature fire extinguisher focusing on the exact spot that needs it. Or maybe they subtly tweak the battery's settings, dialing down the energy flow just enough to keep things under control.

We all want batteries that are powerful, fast-charging, and long-lasting. But none of that matters if they're not safe. The "Cooling the Core" story – the story of phase-change materials and smart thermal management – is the unsung hero of the solid-state battery revolution. It's the foundation that allows us to build a future powered by clean, efficient, and, most importantly, safe energy. It's not just about preventing explosions; it's about building trust. It's about making the future electric, without the fear.

Solid-State Batteries: The Unsung Heroes of the EV Revolution (and Why They Can Take a Punch)

We all love talking about the inside of solid-state batteries (SSBs) – the lithium-metal wizardry that promises to supercharge our electric future. But let's be honest, nobody wants a superhero with a glass jaw. What happens when things get real? I'm not talking about the gentle hum of a lab; I'm talking about the real world: the rogue shopping cart, the unexpected pothole that feels like a lunar crater, the "whoops-a-daisy" moment that makes your heart stop.

That's where the outside of the SSB – the casing – becomes the unsung hero. It's not just a box; it's a high-tech bodyguard for the delicate, energy-packed heart of your EV. And these bodyguards are training hard.

Imagine the National Renewable Energy Laboratory (NREL) as the ultimate dojo for battery casings. They're not pulling any punches. We're talking about impacts that would make a seasoned crash-test dummy wince – forces exceeding 50 G's. To put that in perspective, that's like your EV experiencing several major car crashes simultaneously, and the battery needing to shrug it off.

The secret weapon? It's not brute force; it's grace under pressure. Think of it like this: the old battery casings were like rigid, brittle knights in shining armor. One good whack, and crack – game over. The new generation of SSB casings? They're more like Olympic gymnasts.

These casings are crafted from materials that would make a materials scientist weep with joy. We're talking about polymers that have been hitting the gym, reinforced with carbon fiber, and engineered with the precision of a Swiss watch. They're designed not to resist impact, but to dance with it.

Picture a super-ball. You slam it on the ground, it deforms momentarily, absorbing the energy, and then boing! – it's back to its original shape. That's the principle behind these new casings. They're designed to be flexible, to give way under pressure, like a microscopic shock absorber, and then spring back, protecting the precious battery cells inside.

And who better to learn from than the masters of controlled chaos? The automotive industry, and especially companies like General Motors (GM), are basically the Yoda's of impact resistance. They've been deliberately crashing cars for decades – not out of malice, but out of a deep desire to understand how materials behave when things go boom. Their expertise, gleaned from crumple zones and airbag deployments, is now being poured into making SSB casings tougher than a two-dollar steak.

It's like a beautiful, symbiotic relationship. The nerds in the lab coats (Materials Scientists) are teaming up with the gearheads who love to smash things (automotive engineers). It's a feedback loop of knowledge: smash, analyze, improve, repeat. The goal? To create a battery casing that can not only survive a ridiculous amount of punishment but also prevent the dreaded "thermal runaway" – basically, a battery having a very, very bad day.

Because, at the end of the day, it's all about trust. We need to know that the battery powering our EV isn't just powerful; it's safe. It needs to be able to handle the daily grind, the unexpected bumps, and even the occasional "oh-crap" moment, without turning into a fiery spectacle. It's about having the confidence to hit the road, knowing that your battery's bodyguard has your back. It's about a casing that's built not just to contain energy, but to inspire confidence. Because safety, ultimately, is the coolest feature of all.

Solid-State Batteries: The "No-Boom" Promise, Tested to the Extreme

Imagine your phone battery as a tiny, incredibly energetic toddler. A normal lithium-ion battery (the kind in most phones today) is like that toddler after they've snuck into the candy jar – a hyperactive, potentially explosive situation. Overcharge that battery, and you're basically handing the kid a box of fireworks and a lighter. Things could get messy.

Solid-state batteries (SSBs) are the cool, collected older sibling. They're designed to be inherently safer, even if you accidentally try to stuff them with way more energy than they need. But "designed to be safer" isn't good enough for something that powers your car or keeps your phone alive. We need proof. We need to see that older sibling handle the metaphorical fireworks with calm, collected grace.

That's where the "Proving Grounds" come in – think of it as a battery boot camp, a series of grueling trials designed to push batteries to their absolute limits. And one of the most crucial trials is the dreaded overcharge test.

The Overcharge Torture Chamber (But, Like, a Safe One)

This isn't your average "plug it in and see what happens" scenario. We're talking about deliberately force-feeding the battery way more energy than it's supposed to handle. Imagine trying to cram two Thanksgiving dinners into your stomach – at the same time. That's the level of stress we're talking about. We might pump it up to 150% of its capacity, or even more.

Why the extreme measures? Because in the real world, things go wrong. Chargers malfunction. Software glitches. We need to know that even in these worst-case scenarios, the battery won't turn into a miniature volcano.

Traditional lithium-ion batteries, with their liquid electrolyte "juice," are notoriously bad at handling overcharge. That liquid is flammable, and when things get too hot, it can trigger thermal runaway – a fancy way of saying "uncontrolled fire explosion."

SSBs, on the other hand, are built different. Their "juice" is a solid – think ceramic, a special plastic, or even a sulfur-based compound. It's like comparing a puddle of gasoline (liquid electrolyte) to a brick wall (solid electrolyte). Much harder to set on fire.

The CSI of Battery Testing

During the overcharge test, the battery isn't just sitting there. It's under constant, intense scrutiny. It's like a patient in the ICU, hooked up to every monitor imaginable. Scientists and engineers become battery detectives, meticulously tracking:

    Voltage: Is it spiking like a heart rate monitor in a thriller movie?
    Current: Is the energy flow smooth, or are there dangerous surges?
    Temperature: We're not just taking its temperature in one spot; we're monitoring it everywhere, looking for any hint of overheating. Think of it like a thermal camera, searching for hotspots.
    Gas Evolution: Is the battery "burping" out anything dangerous?

They're looking for any sign of weakness, any crack in the armor. The guiding principles here often come from the Society of Automotive Engineers (SAE) – the rule-makers of the car and plane world. Their standards (like SAE J2464) are like the ultimate battery safety bible, ensuring everyone's playing by the same, incredibly strict rules.

Beyond the Checklist: Audi's Secret Sauce

Companies like Audi, who are betting big on electric vehicles, aren't just ticking boxes. They're going above and beyond. They're taking those SAE standards and adding their own secret blend of torture tests.

Imagine this: they might overcharge the battery while simultaneously blasting it with arctic cold and then desert heat, and shaking it like it's on a bumpy off-road adventure. They might even combine overcharge with other nasty surprises, like poking it with a nail or creating a short circuit. Why? Because they want to simulate every conceivable real-world nightmare to make absolutely sure these batteries are as safe as humanly possible. They need data from as many senarios as they can.

The "Fail Safe" Dream

The perfect outcome? The battery doesn't explode. It doesn't catch fire. Ideally, it doesn't even "fail" in the traditional sense. It might gracefully shut itself down, like a computer going into sleep mode. It might vent a little pressure in a controlled way, like letting a tiny bit of air out of a balloon. But it won't become a hazard.

That's the "no-boom" promise of solid-state batteries. And it's only through these brutal, incredibly detailed overcharge tests – pushing them to the absolute brink – that we can truly trust that promise, and confidently drive, fly, and live in a world powered by the next generation of energy storage.

Subtopic 4: When the "Holy Grail" Bites Back: Lessons from a Solid-State Battery's Fiery Demise

Okay, let's be honest. In the race to build the next-generation battery – the solid-state battery, the "holy grail" of energy storage – sometimes things go kaboom. And in 2022, one of those "kabooms" wasn't just a minor hiccup; it was a full-blown, room-shaking, "Houston, we have a problem" kind of event. We're talking about a solid-state battery (SSB) prototype that decided to demonstrate its potential energy in a way nobody wanted.

Think of SSBs as the sophisticated successors to the lithium-ion batteries in your phone. They promise to hold more power, charge faster, and, crucially, not burst into flames at the drop of a hat. The key is ditching the flammable liquid goo inside current batteries for a solid, stable material. Sounds simple, right? Narrator voice: It wasn't.

The dream is a smooth, super-efficient highway for lithium ions. But what happens when that highway has hidden potholes? Imagine tiny, invisible imperfections – specks of dust, a stray molecule of water, a microscopic rogue particle. These aren't just minor annoyances; they're like throwing a wrench into a finely tuned engine. The ions, instead of cruising smoothly, slam into these obstacles. Friction builds. Heat spikes. And in a battery, uncontrolled heat leads to... well, let's just say it's not pretty.

The 2022 incident wasn't your average lab accident. This was a full-scale forensic investigation. Think CSI: Battery Edition. The team, likely including folks at and referencing the meticulous analysis found in the Failure Analysis journal, became high-tech detectives, poring over the charred remains of the prototype. Their mission: figure out exactly what went wrong.

It wasn't enough to say, "Oh, it was impurities." They needed to know: What kind of impurities? Were they metallic hitchhikers from the manufacturing process? Sneaky moisture that had infiltrated the system? Or maybe unwanted chemical sidekicks created during the electrolyte's birth? This required pulling out the big guns – advanced microscopes and analytical tools that could spot a single rogue atom in a crowd.

The verdict? The solid electrolyte, the supposed bedrock of safety, was riddled with microscopic troublemakers. And the lesson learned was brutal but crystal clear: "Almost pure" isn't good enough. We're talking about a level of purity that would make a cleanroom surgeon look sloppy. This wasn't just about tweaking a process; it was a complete rethink of the entire supply chain. Every raw material, every manufacturing step, even the air in the lab had to be scrutinized.

Solid Power, a major player in the SSB game, has been open about how this fiery setback reshaped their approach. They went from thinking "good enough" to demanding absolute, unwavering purity. It was like going from believing you could build a house on slightly shaky ground to insisting on bedrock foundation carved from a single, flawless diamond.

This explosion, while dramatic, wasn't a death knell for solid-state batteries. It was more like a painful, but incredibly valuable, lesson. It was data delivered in a way that couldn't be ignored – with a bang and a flash. It forced the entire industry to up its game. It underscored that creating the future of energy storage isn't just about brilliant ideas; it's about meticulous execution, relentless testing, and a willingness to learn from even the most spectacular failures.

Beyond Roads: SSB Versatility

The Freedom Battery: Solid-State is Finally Unplugging Us

Remember that sinking feeling? The low-battery icon flashing, your digital lifeline about to flatline? We've all been there, practically chained to a wall socket, rationing our precious screen time like desert travelers guarding water. It's the modern-day equivalent of being stuck in quicksand – the more you move, the faster you sink.

But what if I told you there's a rescue rope on the horizon? Not just any rope, but a super-strong, futuristic one woven from the magic of solid-state batteries (SSBs). Forget incremental upgrades; we're talking about a full-blown jailbreak from battery anxiety.

Imagine this: Your laptop, your trusty workhorse, becomes a wild mustang. You can work a full, intense day, binge-watch your favorite show on a cross-country flight, and still have juice left for a late-night video call with family. We're talking 72 hours of unplugged bliss. Three. Whole. Days. No more frantic searching for outlets, no more awkward power-bank juggling. This isn't a sci-fi movie; this is the near future.

And it's not just about laptops. Companies like Dyson, the folks who made vacuuming almost cool, are betting big on this technology. Their secret battery labs are buzzing with activity, because they see the potential to unleash a cordless revolution. Picture a vacuum that conquers your entire house, multiple times, without needing a recharge. Imagine electric cars that laugh at range anxiety, zipping across states on a single charge, charging up in minutes, not hours.

So, what's the secret? It's like switching from a leaky water balloon to a super-efficient, rock-solid container. Traditional lithium-ion batteries use a liquid electrolyte – think of it as battery "juice" – to shuttle energy around. SSBs ditch the liquid for a solid material. It's a seemingly small change, but it's like upgrading from dial-up to fiber optic.

This solid state unlocks superpowers. First, it's way safer. No more worrying about that rare (but scary) chance of your phone turning into a mini-firework. But the real kicker? Energy density. It's like packing twice the power into the same space, or even shrinking the battery while increasing its lifespan. Think of it as getting double the espresso shots in your morning latte, without making the cup bigger.

The eggheads are already geeking out about it. The journals of IEEE, the folks that practically invented "cutting edge.", are packed with articles on the breakthrough. Top of the list? Better battery life, faster charging, and extended lifespans.

And this isn't just lab-coat fantasies. Companies like Oppo, the smartphone wizards from China, are already grabbing patents for SSB tech like kids grabbing candy. They know this is the key to winning the battery wars, because, who wouldn't want a phone with such capabilities?

Of course, Rome wasn't built in a day, and neither will the SSB revolution. There are still hurdles to jump – making these batteries affordable and mass-producible is a challenge. But the energy (pun intended!) is undeniable. The finish line is in sight.

The era of the perpetually-dying battery is fading. The solid-state revolution is charging up, ready to liberate us from the tyranny of the wall outlet. Are you ready to finally cut the cord and experience true digital freedom? Because it's closer than you ever imagined.

The Electric Sky: How Solid-State Batteries are Making Air Taxis a Reality

Imagine hopping into a silent, electric air taxi for a quick zip across the city, soaring above the traffic jams below. That dream is closer than you think, and it's being powered by something truly revolutionary: the solid-state battery (SSB). Forget clunky, old-school batteries – these are the sleek, super-powered cousins that are about to change how we fly.

Think of it like this: every ounce counts when you're trying to get something off the ground. Traditional batteries are like carrying around a water bottle – useful, but heavy because of all that liquid sloshing around. SSBs are more like a super-concentrated energy bar. They ditch the liquid and use a solid material to move the power, making them much lighter and more stable.

This is where companies like Lilium, with their futuristic-looking eVTOL jets, are making a big bet. They're saying "goodbye" to the extra baggage of traditional batteries. By using SSBs, Lilium expects to slash the weight of their battery packs by a whopping 25%! That's like shedding a whole passenger (and their luggage!) – a game-changer in the world of flight.

What does that extra lightness get you? More whoosh for your buck, basically. Lilium's jets will be able to fly further, carry more people or stuff, and make the whole idea of air taxis way more practical and affordable. Imagine zipping from San Francisco to Napa for a wine tasting in minutes, or skipping the soul-crushing commute on the 405 in Los Angeles.

And Lilium isn't flying solo on this. Joby Aviation, another big name in the flying taxi game, is also on board with SSBs. This isn't just some pie-in-the-sky idea; serious scientific journals, like the Aerospace Science Journal, are publishing research that backs up the awesomeness of SSBs for aviation. It's like the nerds of flight are giving it a big thumbs up!

The real magic of SSBs is that they pack more power into a smaller package. It's like having a tiny, super-strong engine instead of a big, gas-guzzling one. This is especially crucial for eVTOLs, which need a lot of oomph to take off and land vertically, like a helicopter but without all the noise and pollution.

Now, building these super-batteries isn't exactly a walk in the park. Making them at scale, making them consistent, and dealing with some fancier materials – it's all a bit of a challenge. They might cost a bit more upfront, but the long-term payoff is massive: longer flights, safer aircraft, and a much greener way to travel.

Think of companies like Lilium and Joby Aviation as the Wright brothers of electric flight. They're taking the first daring steps, proving that this technology isn't just science fiction. As SSBs get better and cheaper, we'll see them popping up in all sorts of aircraft, big and small. It's the beginning of a true revolution in how we travel, making the skies quieter, cleaner, and a whole lot more accessible. The future of flight? It's electric, and it's solid.

The Tiny Spark: How a Revolution in Batteries is Giving Neurostimulator Patients Their Lives Back

Imagine living with a constant, invisible storm inside you. Chronic pain, the tremors of Parkinson's, the unpredictable seizures of epilepsy – these are battles fought not on a battlefield, but within the intricate wiring of your own nervous system. For many, neurostimulators – tiny electrical marvels implanted in the body – have been a lifeline. They're like miniature conductors, orchestrating the chaotic symphony of nerves back into a harmonious melody. But even lifelines have their limitations.

Think of those early cell phones, the ones that needed charging every few hours. That's kind of what it's been like for neurostimulator patients. The amazing technology that calmed their inner storms relied on batteries – good batteries, but ones with an expiration date. And when that date arrived, it meant surgery. Not a quick pit stop, but a full-blown operation, with all the risks, recovery time, and worry that entails. It was like having to replace the engine in your car every few years – a major disruption, even if the car itself was a lifesaver.

But what if we could change the engine? What if we could make it smaller, safer, and, most importantly, last? That's the revolution Solid-State Batteries (SSBs) are bringing to the world of neurostimulation. Forget the clunky, liquid-filled batteries of the past. SSBs are different. They're like swapping out a leaky, old-fashioned water wheel for a sleek, powerful, and incredibly efficient turbine.

Instead of a liquid "juice" conducting electricity, SSBs use a solid material. It sounds like a small detail, but it's a game-changer. This solid core means the battery is more stable, less likely to cause problems, and – this is the key – it holds its charge for way longer.

Stanford University, a place known for turning "what ifs" into realities, is proving this in their labs. Their research isn't just theoretical; it's showing that SSBs could potentially double the lifespan of these life-changing devices. Double! Imagine cutting those battery-replacement surgeries in half. Instead of a major procedure every five years, maybe it's ten. Or even longer. That's not just less time in the hospital; it's more time living, more time being, without the looming shadow of the next surgery.

This is not some far-off dream. Consider this. Every five years, you are scheduled for surgery, with a week of recovery, then possibly medication to prevent infection. But, if the surgery is only every ten years, the impact is not linear. It is exponential! Five fewer years, five fewer recoveries, five less rounds of medication.

Think about what that means to someone like Sarah, a vibrant artist battling Parkinson's, whose tremors were quieted by a neurostimulator. Before, the thought of another surgery was a constant cloud. Now, with the promise of SSBs, she can focus on her art, on her life, with a newfound freedom. Or imagine David, a father struggling with chronic pain, who can now play with his kids without the constant worry of his device's battery life hanging over him.

Medtronic, a company that's practically synonymous with medical innovation, is already working to bring this future to life. Their neurostimulators are already helping countless people, and by incorporating SSBs, they're poised to make an even bigger difference. It's like giving a superhero an upgraded power source – the same amazing abilities, but with even more endurance and reliability.

And the benefits ripple outwards. Longer-lasting batteries mean engineers can get even more creative. They can design smaller, more sophisticated devices, packed with features that tailor treatment to each individual's unique needs. Imagine a neurostimulator that can "learn" and adapt to your body's rhythms, providing even more precise and personalized relief.

The story of SSBs in medical devices is just unfolding. It's a story of quiet revolutions, of tiny changes making a massive impact. It's a testament to the power of human ingenuity to not just treat illness, but to truly empower people to live fuller, richer lives. It's a story about giving people back the precious gift of time, one tiny, incredibly powerful battery at a time. It is not just about adding years to life, but about adding life to years.

Aloha, Energy! How Solid-State Batteries are Powering Hawaii's Future

Hawaii. The very name whispers of 'āina (land), kai (sea), and lā (sun). But that radiant sun, the lifeblood of these islands, presents a beautiful challenge: how to capture its fleeting brilliance and share it with everyone, even when the 'ōpua (clouds) gather or the hōkū (stars) begin to twinkle. It's a story of energy pono (righteousness, balance) – and the heroes of this story? The unsung guardians of our grids: solid-state batteries (SSBs).

Forget those clunky, old-school batteries. We're talking about the next generation, the kānaka maoli (native Hawaiians) of energy storage – strong, resilient, and perfectly adapted to island life. Imagine them as silent kia'i (guardians), standing watch over Hawaii's precious microgrids. During the day, when the sun blazes with intensity, these SSBs are like tireless workers, diligently absorbing the excess solar energy. That 40% increase in buffering capacity? That's not just a number; it's the difference between a flicker and a steady flame, between dependence and true energy independence.

Picture this: a vibrant 'ohana (family) gathering in a neighborhood powered by a microgrid. Rooftop solar panels are practically singing with energy, capturing the mana (spiritual power) of the sun. In the past, much of that golden energy might have slipped away, lost to inefficiency. But now, nestled discreetly nearby, the SSBs are quietly doing their hana (work). They're like high-tech ipu (gourds), carefully storing the sun's bounty for later use.

What makes these SSBs so special? It's their solid heart. Unlike their liquid-filled cousins, the solid-state electrolyte is like the strong, unwavering core of a koa tree. It gives them incredible energy density – packing more power into a smaller footprint, crucial on islands where space is precious. They're also safer, reducing the risk of those fiery mishaps that nobody wants, and they last longer, a true testament to sustainability.

The respected Energy Storage journal lays out the science, the "why" behind these marvels. But it's companies like Fluence, the real-world kahuna (experts) of energy storage, that are bringing the magic to life. They're not just talking theory; they're installing these systems across Hawaii, turning potential into power.

This 40% increase in buffering? It translates to real-world kōkua (help). It means the lights stay on during a sudden makani pāhili (storm). It means less reliance on wahie (fossil fuels) shipped across vast oceans, bringing Hawaii closer to energy self-sufficiency – a true act of aloha 'āina (love for the land). It means new hana (jobs) for local communities, building a greener economy. It's about keeping our homes comfortable, our hospitals running, our schools bright, and our vibrant tourism thriving – all powered by the clean, reliable energy stored within these silent guardians.

And even when the sun dips below the horizon, painting the sky with fiery hues of orange and purple, the SSBs are there. Homes and business continue operation, with clean renewable energy.

These SSBs aren't just storing kilowatt-hours; they're storing the promise of a brighter future. They're a testament to human ingenuity, a perfect hui (partnership) between cutting-edge technology and the ancient wisdom of living in harmony with nature. They represent the true spirit of aloha – a commitment to sustainability, resilience, and a future where the sun's energy powers a thriving, vibrant Hawaii.

Rules of the Road: Regulatory Realms

Okay, let's ditch the formal textbook vibe and talk about EV safety like we're chatting over coffee, especially that hot topic of solid-state batteries (SSBs) and how we make sure they don't, you know, explode.

Think of UNECE R100 as the ultimate EV safety bible. It's like the UN's way of saying, "Hey, let's make sure these electric cars don't turn into unexpected fireworks." It's not just for Europe, though – it's a big deal pretty much everywhere. It's been updated a few times, and these updates matter a lot for new battery tech.

Now, solid-state batteries? They're the rockstars of the future battery world. Imagine your phone battery, but way, way better. They promise to hold more power, charge super-fast, and – crucially – be a lot safer. Traditional EV batteries have this liquid goo inside (the electrolyte), which can be a bit...flammable. SSBs swap that out for a solid material, which is inherently less likely to go boom. Think of it like switching from a leaky gas can to a solid brick of fuel – the brick is much less likely to cause a problem.

But – and this is a big "but" – new doesn't always mean perfect. We still need to make absolutely sure these new batteries are safe. That's where our "safety bible," R100, comes in.

R100 doesn't care if it's a classic battery or a fancy new SSB. It puts all batteries through a gauntlet of torture tests. Seriously, it's like battery boot camp. We're talking:

The Shake, Rattle, and Roll: Imagine your car vibrating like crazy on a bumpy road. R100 simulates that to make sure the battery doesn't fall apart.

The Hot and Cold Show: Batteries hate extreme temperatures. R100 throws them into ovens and freezers to see how they cope.

The Short Circuit Surprise: Oops! What happens if the wires get crossed? R100 checks that the battery doesn't freak out.

The Crash Test Dummy Treatment: Because, well, accidents happen. R100 literally crushes batteries to see if they stay contained.

The No-Shock Zone: Making sure you can't get zapped.

The No-Over-Doing-It Zone: Making sure the battery can't be over charged, or over-discharged.

This isn't just some theoretical stuff. These tests are meant to mimic what your EV might actually go through in its lifetime, and then some.

So, where does the "real world" come into this? Let's take a trip to Japan. They're like the EV ninjas – always at the forefront of car tech. Japan has basically said, "Yep, R100, we're on board." They've taken this UN rulebook and made it part of their own national laws.

This isn't just about following the rules. It's about making sure that any EV you buy in Japan is super safe, meeting standards recognized around the globe. And it makes it easier for Japanese car companies to sell their EVs everywhere else, because everyone's playing by the same safety rules. Plus, it gives you, the driver, a warm, fuzzy feeling knowing your car isn't going to turn into a science experiment gone wrong. The research in Transport Policy is a good place to go if you're into the nitty-gritty details of how this all works out.

For Solid State Batteries, it is a work in progress. R100, and similar rules around the world, will probably need some tweaks. Think of it like updating your phone's operating system – you need to make sure the new software works with the new hardware. The rule-makers, the testers, and the battery builders will all work together on this.

The bottom line? Making EVs safe, especially with these fancy new batteries, is a team effort. It's like a global collaboration to make sure we can all enjoy the benefits of electric cars without the scary bits. It is always changing, always getting better. It is a pretty cool process, if you think about it.

The Solid-State Battery Gold Rush: Where Fortunes are Forged in Fire (and Filing Cabinets)

Forget pickaxes and dusty claims – the real gold rush of the 21st century is happening in sterile labs and the hallowed halls of patent offices. We're talking about solid-state batteries (SSBs), the holy grail of energy storage, and the battle to own them is fierce. It's a modern-day land grab, but instead of acres, it's intellectual territory, mapped out in a staggering 10,000+ patents (and counting!).

Imagine a vast, unexplored wilderness, brimming with the promise of untold riches. That's the SSB landscape. And every patent is a flag planted, a stake driven into the ground, claiming a piece of this revolutionary technology. Leading the expedition? Toyota, the grizzled prospector with over 1,300 patent filings to its name. They're clearly building an empire, one patent at a time.

But this ain't a solo adventure. This is a free-for-all, a chaotic scramble with players like CATL – who are currently locked in a legal brawl that feels more like a multi-continent cage match than a patent dispute. Lawsuits flying like punches, counter-claims echoing in the courts... it's all part of the brutal, beautiful game of protecting your turf and trying to knock your rivals off their claims. Why all the drama? Because owning the right patent in this space is like striking the mother lode – it could mean billions.

It is a hard reality that patent ownership isn't a finish line, it is barely the starting block. You must then maintain and enforce the ownership.

The sheer volume of patents is mind-boggling. It's like a blizzard of paperwork, each document representing a company's desperate search for the "secret sauce" – that one breakthrough that will make their SSB the best, the fastest, the most powerful. Maybe it's a magical new electrolyte, a mind-bending manufacturing trick, or a cell design so elegant it makes other batteries look like clunky toys. Whatever it is, someone, somewhere, is trying to lock it down with a patent.

It really is all about defining the space, and defending it.

These companies aren't just grabbing patents at random; they're building fortresses. Think of it like assembling a massive, interlocking puzzle of intellectual property. Each patent is a brick, carefully placed to create an impenetrable wall around their core technology. It's a defensive strategy worthy of a medieval warlord, designed to keep competitors at bay.

Navigating this legal labyrinth, as journals like World Patent Information can attest, is like threading a needle in a hurricane. You need lawyers who are part patent law gurus, part tech whisperers, and part bloodhounds, able to sniff out potential threats and opportunities hidden within the dense thicket of claims. One wrong move, one missed detail, and you could be facing financial ruin or, worse, be shut out of the game entirely.

It is not for the faint of heart.

But here's the thing: these "Patent Wars" aren't just about legal squabbles. They're a symptom of something amazing. They're proof that incredible innovation is happening right now. Each patent, each lawsuit, each hard-fought victory represents a step towards a future powered by batteries that are safer, more powerful, and longer-lasting than we ever thought possible.

So, while the courtroom battles might seem messy and expensive, they're a sign of a revolution in progress. It's a high-stakes game, a wild west of innovation, and the future of energy is being forged, one patent – and one legal showdown – at a time. The fight is for the very future of power.

The Great Australian Lithium Squeeze: Will SSBs Go Flat?

Imagine a bustling port in Western Australia. Cranes swing, loading massive containers of what looks like ordinary rock – spodumene concentrate – onto cargo ships. This isn't just rock; it's the key ingredient in the next generation of batteries, the solid-state battery (SSB), the thing that's supposed to make electric vehicles (EVs) go further, charge faster, and be safer. It's the future, and right now, a huge chunk of that future – almost half of the world's mined lithium, a whopping 46.8% – comes from right here.

But there's a change in the air. You can almost smell it, a whiff of...well, national ambition. Australia, blessed with enough lithium reserves (we're talking an estimated 6.2

million metric tons) to power a lot of EVs, isn't content being just a quarry for the world. They've got a vision, and it's painted in the colors of high-tech factories, skilled jobs, and a "Made in Australia" stamp on finished batteries, not just raw materials.

Meet Sarah, a fictional (but representative) geologist working at a lithium mine in the Pilbara. She's proud of the work she does, but she's also heard the whispers. The government's talking about "value-adding," about keeping more of this "white gold" at home. It's not about slamming the door shut on exports – the billions of dollars those exports bring in (around AUD 20 billion in the last financial year!) are too important. It's more like…turning the tap down a bit.

Think of it like this: Australia's got a delicious cake (the lithium), and instead of selling everyone slices of raw batter, they want to bake the whole cake themselves, maybe even add some fancy icing (battery manufacturing). They're not saying "no cake for you!" to the rest of the world, but they're definitely saying, "We're keeping a bigger piece for ourselves."

This "bigger piece" comes in the form of government funding – think hundreds of millions of dollars in grants – for companies willing to build processing plants and battery factories down under. It's like a golden handshake, saying, "Stay here, build here, and we'll help you out." There's also that slightly mysterious phrase, "national interest," hanging in the air. It means that if a big overseas company wants to buy up an Australian lithium mine, the government's going to take a very close look to make sure it fits with their grand plan.

Now, zoom across the ocean to a gleaming SSB factory in South Korea. The CEO, let's call him Mr. Park, is pacing. His supply chain, a finely tuned machine designed to bring in Australian lithium, is suddenly facing headwinds. He's not facing an outright ban, but he is feeling the pressure. He's seeing the writing on the wall. He needs a Plan B, a Plan C, maybe even a Plan D.

Mr. Park, and countless others like him around the world, are starting to look elsewhere. They're making deals in South America, where Chile (producing about 30% of the world's lithium) has its own way of doing things, with hefty royalties (up to 40% in some cases!). They're exploring mines in Africa, even investing heavily in battery recycling – anything to avoid being too reliant on one source.

The price of lithium? It's been on a rollercoaster. Think peaks and valleys, highs of over $8,000 per tonne for spodumene, then crashing down. It's enough to give any CEO a serious headache. It's not all Australia's fault, but the uncertainty about their future policies definitely isn't helping.

The big question, the one Sarah, Mr. Park, and everyone in the industry is asking, is this: Will Australia's ambition help create a thriving domestic battery industry, or will it accidentally slow down the global SSB revolution? Will the squeeze on lithium supply force innovation, pushing companies to find new battery chemistries, or will it just delay the electric future we've all been promised?

There is no way to know. It is a giant experement, and we are watching it play out in real time.

It's a global chess game, with lithium as the queen, and Australia's making some bold moves. Whether it's checkmate or a stalemate remains to be seen. But one thing's for sure: the battery world is watching, and the stakes are incredibly high.

India's Electric Vehicle Push: Less Carrot, More Stick (But It's Working!)

India's air isn't cleaning itself, and let's be honest, the sudden surge in electric vehicles (EVs) isn't just because everyone suddenly became eco-warriors. The real hero (or villain, depending on how you see it) is a hefty dose of government regulation, dressed up in the rather unsexy name of Bharat Stage VI (BS-VI) emission norms. Think of it as the government's not-so-subtle way of saying, "ICE age is over, folks! Time to plug in."

We were seriously behind the curve. Before BS-VI, India was chugging along with outdated emission standards (BS-IV), letting our cars and bikes happily pollute away. Then, in a move that felt a bit like skipping a grade in school, we jumped straight to BS-VI. This wasn't a minor tweak; it was a major overhaul. Car companies had to practically rebuild their engines from the ground up, adding all sorts of fancy (and expensive) tech to make them less polluting.

And that's the clever part. This regulatory "stick" made traditional gasoline and diesel cars significantly pricier. Suddenly, that shiny new EV, which used to seem like a luxury, started looking a lot more reasonable. The government didn't just level the playing field; they tilted it towards electric. It's like making vegetables cheaper than candy – a nudge in the right direction, even if it's not entirely voluntary.

Nowhere is this playing out more dramatically than in Delhi. The capital, famous for its smog so thick you could practically taste it, decided to double down on the EV push. Think of Delhi as the enthusiastic student who's not only doing the assigned homework (BS-VI) but also volunteering for extra credit.

Delhi's EV policy is like an all-you-can-eat buffet of incentives. We're talking big discounts on the purchase price, zero road tax, and free registration. Plus, they're building charging stations like there's no tomorrow, tackling that nagging fear of running out of juice mid-commute (aka "range anxiety"). It's the national stick (BS-VI) combined with the local carrot (Delhi's incentives) – a powerful one-two punch.

Is this a bit... forceful? Maybe. Does it feel a little like being told what's good for you? Sure. But is it working? You bet. Look around – EVs are zipping around Delhi, and it's not just because of a sudden surge in environmental consciousness. It's because the government made the green choice the smart choice, the affordable choice.

Of course, we're not at the finish line yet. We still need charging stations everywhere, a power grid that can handle the load, and a way to deal with all those old batteries responsibly. But the ball is rolling, and it's gaining speed. BS-VI was the shove, and cities like Delhi are providing the momentum.

It's a testament to the power of policy. Sometimes, you need a little (or a lot) of government muscle to get things moving. It proves that strong environmental rules, combined with practical support, can force a change, making the sustainable option the logical one, too. India's electric revolution isn't just happening; it's being engineered, one regulation at a time. And, frankly, it's pretty electrifying to watch.

Minds at Work: R&D Ecosystems

From Dusty Tomes to Dusty Roads: Oxford Cracks the Code for Batteries That Actually Last

Forget the stereotype of universities as places where professors only debate the meaning of life in Latin. At the University of Oxford, they're tackling a problem that's far more pressing for most of us: the agony of a constantly dying phone battery and the "range anxiety" that plagues electric car owners. And they're doing it with something that sounds like it belongs in a fantasy novel: garnet.

No, not the kind you wear on a ring. We're talking about a special kind of garnet that's the key ingredient in the next generation of batteries – solid-state batteries, or SSBs. These aren't your grandpa's leaky, fire-prone batteries. SSBs are the superheroes of the battery world, promising to make electric cars go further, safer, and without the constant fear of running out of juice mid-highway.

The catch? Until now, SSBs have been like that brilliant but flaky friend – full of potential, but unreliable. They tended to die young, making them about as useful as a chocolate teapot for everyday use.

Enter the Oxford team. Imagine them not in stuffy lecture halls, but more like a team of battery detectives, hunched over microscopes, peering into the atomic world of garnet. They're not just mixing chemicals; they're like alchemists, transforming this material into something truly extraordinary.

Their secret? They've figured out how to make the garnet "highway" inside the battery – the electrolyte that lets the electricity-creating ions zip back and forth – super smooth and super durable. Think of it like upgrading a bumpy dirt road to a pristine, eight-lane superhighway. The result, as published in the seriously impressive journal Nature Materials, is mind-blowing: they've doubled the lifespan of these garnet-powered SSBs.

Doubled! That's not a tiny tweak; that's like going from a horse-drawn carriage to a spaceship in terms of battery performance. Picture this: Your electric car could potentially travel twice as far on a single charge. Your phone? Maybe you'd only need to charge it every other day, and it would consistently work. No more frantic searches for a power outlet!

And this isn't just some pie-in-the-sky academic dream. This is real-world, get-your-hands-dirty science. Jaguar Land Rover, the folks who make those sleek electric cars, are deeply involved. They're not just throwing money at the problem; they're working hand-in-hand with the Oxford researchers, turning this lab magic into something you might actually drive in a few years.

It's like a beautiful collaboration: the brainpower of academia merging with the practical know-how of industry. The Oxford team isn't just building a better battery; they're obsessively studying why it's better. They're like battery whisperers, understanding the secret language of ions and materials, learning how to make this garnet highway even smoother and more resilient.

Of course, there's still a long road ahead. Turning a lab breakthrough into something you can buy at the store is a marathon, not a sprint. There are challenges in making these batteries on a massive scale, keeping them affordable, and making sure they're as reliable as your trusty old toaster.

But the Oxford research is a massive leap forward. It's proof that the "ivory tower" can be a powerhouse of innovation, that brilliant minds can solve real-world problems, and that the future of batteries – and maybe even the future of how we power our world – might just be found in a little bit of perfectly engineered garnet. It's a reminder that sometimes, the most exciting discoveries happen where you least expect them, in the pursuit to eliminate those first-world modern anxieties.

Beyond the Hype Cycle: Where Big Companies Make Real Moonshots

Forget the image of a lone genius tinkering in a garage. Some of the most audacious leaps in technology aren't happening in Silicon Valley lofts or university basements – they're brewing inside the sprawling, often secretive, research labs of corporate giants. These aren't places where they just tweak next year's model; they're where companies gamble on the far future, sometimes with stakes that would make a casino owner sweat.

Consider Panasonic. They're not just fiddling with better battery packs. They're going all in on solid-state batteries (SSBs) – a technology that sounds like something out of science fiction, but could revolutionize everything. We're talking a mind-blowing $4 billion bet. That's not "let's see what happens" money; that's "we're staking a claim on the future of power" money. Imagine electric cars that can drive across the country on a single charge, or homes powered by energy storage that's as safe as a brick. Panasonic is aiming to make this a reality by 2029.

Why this colossal gamble on something that's still pretty experimental? Because the potential reward is astronomical. SSBs are like lithium-ion batteries on steroids – more power, smaller size, and way less likely to, you know, explode. For a company like Panasonic, which makes everything from electric car parts to home appliances, mastering SSBs is like finding the Holy Grail of energy.

But here's the secret sauce: it's not just about the money. You can't just throw billions at a problem and expect magic to happen. Innovation is a messy, human endeavor. It needs brilliant minds, sure, but it also needs the right environment. Think of it like a jazz band – you need talented musicians, but they also need the freedom to improvise, riff off each other, and even hit a few wrong notes along the way. R&D Management, the journal that's like the bible for innovation nerds, constantly emphasizes this: it's about the people and how they work together, not just the lab equipment.

Nissan's research labs are a perfect example. They're not just making next year's sedans a little shinier. They're the folks dreaming up self-driving cars and materials that sound like they belong in a superhero movie. They have dedicated teams working on stuff that might not even become a product for a decade, if ever. That's how you stay ahead of the game – by constantly looking over the horizon.

Corporate labs are these fascinating, almost contradictory places. They have to deliver results that keep the shareholders happy, while also funding wild, long-shot projects that might fizzle out completely. The best ones – the ones that truly change the world – find a way to walk that tightrope. They create space for "what if?" thinking, but they always keep one eye on the company's big-picture goals.

Panasonic's $4 billion SSB investment is important. It's more than a product roadmap. It's showing a company who's

willing to shape it, instead of waiting for the future, placing their marker down, and take a risk.
And they're doing it not in some far-off, mythical lab, but right there, within their own walls. That takes guts, vision, and a whole lot of faith in the power of human ingenuity.

Canada's $500 Million Bet on the Electric Avenue: It's More Than Just Money

Picture this: the world's biggest automakers are in a Formula 1 race, but instead of roaring engines, it's the silent hum of electric motors. Canada? We're not just watching from the stands; we're aiming to be in the pit crew, supplying the crucial parts that make those electric speed demons go. And we're putting serious money on the table – a cool $500 million – to make it happen.

The star of our pit crew? Graphite anodes. Sounds a bit technical, right? Think of them as the unsung heroes inside those lithium-ion batteries that power electric vehicles. Without them, your Tesla's not going anywhere. And Canada's realizing that being a leader in the EV revolution means more than just buying the finished cars; it means controlling the ingredients that make them run.

This isn't just about going green, though that's a huge part of it. It's about smart economics. We're not content with just digging up the graphite (which, by the way, we have plenty of) and shipping it off. We want to build the whole battery ecosystem right here. Imagine a "Silicon Valley of EVs" taking root, creating good-paying jobs, attracting brilliant minds, and making Canada a global player in this booming industry.

The $500 million isn't a blank check; it's seed money. It is like venture capital, but not for a profit only. The government's playing the role of a savvy investor, making it less risky for private companies to jump in. They're saying, "Hey, we'll help you get started, we'll share the initial risk, so you can build the future of EV tech here."

And where's "here"? Places like Ontario are buzzing with potential. They've got the car-making history, the tech-savvy workforce, and now, thanks to this fund, a real shot at becoming a major EV hub. We're already seeing the sparks fly – research labs humming, startups launching, and partnerships forming. It's like watching a new industry being born, right before our eyes.

This strategy is not a new concept. Prestigious journals, like Science Policy, regularly publish articles which talk about how nations are all trying to be on top of the next technological wave.

The government isn't trying to run the whole show. They're more like a conductor, bringing together universities, research labs, and private companies to create a symphony of innovation. They provide the funding, set the stage with smart policies, and let the creativity flow. And funding will go to the basic research.

But here's the really cool part: it's not just about graphite anodes. It's about everything that surrounds them. Think of it like planting a tree – you don't just get the tree, you get the shade, the birds that nest in it, the whole ecosystem that springs up around it. This investment could spark growth in materials science, battery manufacturing, charging stations, and even the software that makes batteries smarter.

It's a gamble, sure. Any big, bold move is. But imagine the payoff: a cleaner planet, a stronger economy, and Canada right at the forefront of the EV revolution. It's a marathon, not a sprint, but with this $500 million boost, we're definitely off to a running start. It is a long-term strategy, and every marathon begins with a first step. We removed the policy talk, because it isn't as catchy, and focused on results.

Beyond the Lone Wolf: Battery500's Quest for the Solid-State Battery Holy Grail

Forget the image of a solitary scientist toiling away in a dimly lit lab. Battery500 isn't about isolated "eureka" moments; it's about a scientific party, a vibrant, buzzing hub where the best minds in battery tech are throwing ideas around like confetti. This isn't your grandpa's research project – it's a 50+ strong super-team, on a mission to conquer the notoriously tricky solid-state battery (SSB).

Imagine a bustling marketplace, not of spices or silks, but of knowledge. Universities, research powerhouses, and even industry rivals have set aside their competitive instincts (at least for now!) and joined forces. Why? Because SSBs are the Mount Everest of battery technology. They're incredibly promising, but also incredibly complex. No single climber, no matter how brilliant, can conquer this peak alone.

Battery500 is like a scientific orchestra. You've got the materials scientists, the virtuosos of the atomic world, fiddling with the very building blocks of the battery. Then there are the battery manufacturers, the seasoned conductors, bringing their real-world experience – the "we've tried that, and it blew up" kind of wisdom. And to tie it all together, you have the modeling gurus, the digital prophets, simulating battery life in ways that would make your head spin.

This isn't just some kumbaya circle of scientists holding hands. This open-source approach is backed by hard evidence. Smart folks in lab coats, writing in fancy journals like Research Policy, have been saying for years that collaboration is king, especially when facing down scientific dragons like SSBs. Battery500 is basically a live-action demonstration of that principle – a "proof is in the pudding" kind of deal.

It is similar in concept to the EU's big-picture, "let's-revolutionize-batteries" project, Battery 2030+. Think of Battery 2030+ as the visionary architect, drawing up the grand plans for a battery-powered future. Battery500 is the team of skilled builders, down in the dirt, actually laying the foundation, brick by painstaking brick.

The benefits of this brain trust are huge. First, it's like a giant "been there, done that" database. No more wasted time and money chasing solutions that someone else already proved were duds. Second, it's a breeding ground for unexpected breakthroughs. A chemist's casual observation might be the aha! moment a battery designer desperately needed. It's the magic that happens when you get diverse minds rubbing elbows. Third, it reduces the intense, sometimes stifling pressure to be first to patent, first to market. This pressure cooker is taken off of the flame, so that the best tasting meal, or, battery, can be prepared.

Now, let's be real: wrangling this many brilliant minds isn't always a walk in the park. Figuring out who owns what idea (intellectual property, they call it) is like untangling a giant ball of Christmas lights. Keeping everyone on the same page, communicating effectively, is like herding cats... very, very smart cats.

But Battery500 is making it work, and doing so with a measure of elegance. They're showing the world that the old-school, "my secrets are my own" approach is about as useful as a rotary phone in the age of smartphones. In the quest for the next generation of batteries – and probably in many other fields – the future looks a lot like Battery500: open, collaborative, and powered by the combined brilliance of many, rather than the genius of a few. They're not just building better batteries; they're building a better way to do science. The collective isn't just stronger; it's unstoppable.

Powering Up: Infrastructure Innovations

Charging Constellations: Plugging into the Future of Driving – A Road Trip, Not a Revolution

Forget sterile talk of "infrastructure" – let's talk about freedom. Imagine a road trip, the windows down, the music up, but instead of the rumble of a gas engine, you hear... almost nothing. Just the quiet hum of electric power, carrying you across continents, fueled by a growing network of charging stations that are more like friendly oases than clunky gas pumps. This isn't some distant sci-fi dream; it's happening now.

We're witnessing the birth of electric mobility's nervous system – a vast, interconnected web of charging points that's spreading across the globe like wildfire. We've passed the half-million mark for public chargers worldwide – that's 500,000 little beacons of hope, each one saying, "Go ahead, explore. We've got you covered."

But this isn't just about plugging in your car; it's about plugging into a lifestyle. Remember the early days of the internet? Dial-up was exciting, but broadband was life-changing. We're at that broadband moment for EV charging. It's not enough to just have a charger; we need fast chargers, the kind that let you top up your battery while you grab a coffee and check your messages. Not hours, but minutes.

Think of companies like ChargePoint as the architects of this new electric landscape. Their focus on 150 kW hubs isn't just about technical specs; it's about erasing "range anxiety" – that nagging little voice in the back of your head that whispers, "What if I run out of power?" These hubs are like the super-powered rest stops of the future, strategically placed to make long-distance electric travel a breeze. They are not only the pit stops, but places of comfort.

This explosive growth isn't magic; it's the result of smart planning and a real desire for change. Smart people doing research, as shown in places like Transport Reviews, show that the more chargers there are, and the easier to use they, are the more people will switch to electric. Basic logic, the type that is going to push us forward.

And governments are getting in on the act, too. Take the U.S. National Electric Vehicle Infrastructure (NEVI) program, for example. This isn't just about throwing money at chargers; it's about building a smart network. It's like laying down the train tracks before the train arrives, making sure that fast chargers are strategically placed along highways, so you can cruise from coast to coast without a second thought. They are making sure the infastructure is built, and built correctly.

The NEVI program is like a masterclass in forward-thinking. It's not just reacting to the present; it's anticipating the future. It's about making electric road trips not just feasible, but genuinely enjoyable – something you'd choose to do, not something you'd have to compromise on.

This isn't just about technology; it's about a fundamental shift in how we move. It's about connecting cities and towns with clean energy, empowering us to explore without leaving a heavy footprint. It's a big puzzle, and it takes everyone – governments, businesses, bright minds in research labs – working together to solve it.

But the picture is becoming clearer every day. The future of driving is electric, and that future is bright. These charging "constellations" are lighting up, one by one, creating a network that's not just powerful, but empowering. It's a future where the open road is truly open, and the journey is just as exciting as the destination. Get ready to plug in and go!

From Pit Stop to Swip Stop: The Electric Revolution's Need for Speed

Let's face it: electric vehicles are awesome. They're quiet, they're zippy, and they're good for the planet. But that charging time? Oof. It's like waiting for dial-up internet in the age of fiber optics – a total buzzkill. But what if I told you there's a way to "refuel" your EV faster than you can say "supercharged"? Enter battery swapping, the pit-crew-inspired solution that's turning the EV world on its head.

Forget plugging in and scrolling through your phone for half an hour. Imagine pulling up to a station, and – whoosh – a robotic arm (or a super-efficient human) swaps out your drained battery for a fully juiced one. We're talking Formula 1 levels of speed, here, folks. And the undisputed champion of this speedy revolution? Gogoro.

This Taiwanese powerhouse has basically turned Taiwan into a battery-swapping paradise. With over 2,500 GoStations scattered across the island, finding a fresh battery is easier than finding a bubble tea shop (and that's saying something!). These aren't your clunky, industrial charging stations, either. Think sleek, strategically placed hubs that blend seamlessly into the urban landscape. And the swap itself? Six seconds. Six. You'll spend more time adjusting your mirrors than you will waiting for power.

This isn't just a cool tech demo; it's a game-changer for anyone who's ever felt that creeping "range anxiety." That little voice in your head that whispers, "Are you sure you have enough juice to get there?" A study in the Journal of

Cleaner Production even confirmed it: battery swapping melts away that fear, making EVs feel as reliable as their gas-guzzling counterparts. Because let's be real, nobody wants to be that person stuck on the side of the road with a dead scooter, especially in the heart of a city like Taipei.

And the battery-swapping love is spreading! India, the land of a billion scooters (okay, maybe not quite a billion, but you get the idea), is jumping on the bandwagon. Ola Electric, a major player in the Indian EV scene, is piloting its own swapping programs. In crowded cities where two-wheelers rule the road, this is pure genius. It's like giving every scooter an instant energy boost, whenever and wherever they need it.

But here's where it gets even better. Battery swapping isn't just about speed; it's about future-proofing your ride. Think of it like this: you buy your EV, but you're not married to its battery forever. As battery tech gets better (and it always gets better), the swapping network upgrades, and you get the benefits without buying a whole new vehicle. It's like getting a constant stream of performance upgrades, delivered straight to your ride.

And there's more! These swapping stations are like secret superheroes for the power grid. They can charge up those batteries during off-peak hours, when electricity is cheaper and more plentiful. This helps balance the load, making the whole energy system more efficient and even paving the way for more renewable energy. It's a win-win-win: faster refills for you, a smoother grid for everyone, and a greener planet.

Now, I'm not going to lie, there are a few bumps in the road. The biggest one? We need everyone to play nice. Imagine if every gas station only worked with one specific car model – total chaos, right? We need some standardization in the battery world, so different EVs can share the swapping love. But the potential rewards are so huge, it's a challenge worth tackling head-on.

Battery swapping is more than just a convenience; it's a whole new way of thinking about energy and how we move around. It's about creating a seamless, sustainable, and seriously fast ecosystem that'll make electric vehicles the obvious choice for everyone. With pioneers like Gogoro and Ola showing us the way, the future of "refueling" is looking lightning-fast and incredibly exciting. Get ready for the swip stop revolution!

Your EV: Not Just a Car, But a Dancing Partner in the Energy Revolution

Forget everything you think you know about electric vehicles. Sure, they're sleek, silent, and good for the planet. But they're about to become something far more extraordinary: partners in a revolutionary energy dance. Imagine your EV not just as a way to escape traffic, but as a secret weapon against blackouts, a mini power station on wheels, and maybe even a way to earn a little extra cash.

California, the land of sunshine and innovation, is already leading the charge (pun intended!). They're experimenting with something called Vehicle-to-Grid (V2G) technology, and it's turning the energy world on its head. Think of it like this: your EV's battery isn't just a fuel tank; it's a two-way energy superhighway.

Picture this: It's a scorching summer afternoon. Everyone's blasting their AC, and the power grid is groaning under the strain. Suddenly, your EV, plugged into your smart charger, springs to life. It's not taking power; it's giving it back. A small contribution, maybe 10% of its battery capacity, but multiply that by thousands, even millions, of cars, and you've got a powerful, responsive energy reserve.

It's a beautifully choreographed "Grid Dance." The grid, like a conductor leading an orchestra, signals a need for energy. Your EV, along with its electric brethren, responds instantly, a perfectly synchronized chorus of electrons flowing back into the system. No more fear of flickering lights or sudden darkness – your car is helping to keep the lights on for everyone.

This isn't some far-off, sci-fi dream. The brainy folks at Smart Grid journal are already mapping out the choreography, figuring out the intricate steps and rhythms of this energy ballet. Companies like Enel X aren't just talking about it; they're building the stage – installing the smart chargers, writing the software, and working with car manufacturers to make sure every new EV is ready to hit the dance floor.

What's in it for you? Well, imagine getting paid for helping to power your neighborhood. Utilities might actually reward you for sharing your EV's energy. Or picture this: a storm knocks out the power, but your home stays lit, your fridge stays cold, all thanks to your car. Forget those clunky, gas-guzzling generators; your EV is your silent, clean, and surprisingly powerful backup power source.

What's in it for all of us? A more stable, reliable, and resilient power grid. As we embrace the beautiful, but sometimes unpredictable, energy of the sun and wind, V2G acts like a giant shock absorber, smoothing out the bumps and ensuring a steady flow of power.

V2G is more than just clever technology; it's a fundamental shift in how we think about energy. It's about turning every EV owner into an "energy prosumer" – someone who both consumes and produces power. It's about blurring the lines between "us" (the energy users) and "them" (the power companies). It's about creating a collaborative, interconnected energy ecosystem, where every car is a potential power player.

The Grid Dance is just getting started, and it's a performance you won't want to miss. It's a revolution on wheels, and it's powered by the very cars we'll be driving tomorrow. It's a new movement, and it's electrifying.

Singapore's Electric Dream: Plugging into a Quieter, Cleaner Tomorrow

Singapore. The name conjures images of gleaming skyscrapers, lush gardens, and a relentless pursuit of the future. But beneath the polished surface, a quiet revolution is brewing – a shift away from the rumble of combustion engines to the gentle hum of electric vehicles. This isn't just a trend; it's a meticulously planned urban metamorphosis, and its beating heart is a truly audacious goal: 60,000 EV charging points by 2030.

Imagine 60,000 tiny power oases, woven into the very fabric of the city. This isn't about slapping a few chargers in the corner of a parking garage. This is about making EV charging as commonplace as finding your favorite char kway teow stall – always there, always ready. It's about fundamentally changing the way energy flows through Singapore, like upgrading the city's circulatory system.

This isn't some pie-in-the-sky dream. Top urban planning researchers have crunched the numbers, looked at the projected growth of EVs, and considered the unique, space-conscious realities of this island nation. 60,000 isn't just a nice round number; it's the right number. And Singapore isn't alone. Far across the globe, in the cycling-friendly streets of Copenhagen, a similar spirit of electric ambition is taking hold. While the details differ, the core idea is the same: make charging so easy, so seamless, that it becomes second nature. Sharing their methods of application is one key to making this transition a success.

Let's paint a picture of this electric future. You're cruising through Singapore in your EV – a silent, stylish extension of your personality. Need a boost? Forget frantic searches for a dedicated charging station and the dread of a long wait. You simply pull up to your HDB flat. Your block, like many others, has embraced the electric future, boasting a suite of charging points. Or maybe you're at work. The office carpark? It's not just a place to park; it's a smart energy hub, cleverly balancing power distribution so everyone gets the juice they need without causing a city-wide blackout.

But this is about more than just convenience; it's about building a whole new relationship between people, their cars, and their city. Think slick apps that act like your personal charging concierge, showing you available spots in real-time and offering dynamic pricing. Imagine even using your EV as a mobile power bank, feeding energy back into the grid during peak demand and earning credits – a true win-win. It is a practical symbiosis.

Of course, there will be bumps in the road. How do you retrofit older buildings without turning them into construction zones? How do you beef up the power grid to handle the surge in demand without, well, causing surges? And, perhaps most importantly, how do you convince everyday Singaporeans that "range anxiety" is a relic of the gasoline age?
That is where buy in is most crucial.

These are real challenges, requiring ingenious engineering, smart technology, and a healthy dose of public education. But the prize? It's breathtaking. Imagine a Singapore where the air is noticeably cleaner, where the constant background hum of traffic is replaced by a softer, more peaceful soundscape. Imagine a city that's not just surviving the 21st century, but thriving in it, powered by clean energy and driven by a collective commitment to a greener future.

Singapore's 60,000 charging point target isn't just a statistic; it's a declaration of intent. It's a beacon, illuminating a path towards a future where our cities are not just smart, but sustainable. It's a future that feels tantalizingly close, a future where the quiet hum of electric vehicles is the soundtrack of urban life. And that future will require everyone playing their part, the mechanics, to the politicians, to the car makers.

Culture in Charge: Societal Shifts

The Electric Hum: Are We Finally Ready to Ditch the Dinosaurs?

Remember that feeling? That gritty, cough-inducing cloud of exhaust you'd get stuck behind at a red light? Yeah, that one. For years, it felt like the soundtrack to our lives, the unavoidable price of getting from A to B. Electric cars? Those were for Elon Musk and people who hugged trees for a living (no offense to tree-huggers, we love you!). They were a whisper of a future, not the roar of the present.

But hold on a second. Something's...different. It's like a low, electric hum is starting to replace that dinosaur roar. And it's getting louder.

Europe, that trendsetter across the pond, is practically shouting it from the rooftops. We're talking about EV acceptance hitting a jaw-dropping 70% in 2023! Seventy percent. That's not your eccentric aunt Mildred anymore; that's your dad, your best friend, maybe even that guy who always parks his gas-guzzler diagonally across two spaces. (We all know one.)

So, what flipped the switch? (Pun intended, couldn't resist). It's not some overnight miracle; it's more like a slow-burning realization that finally caught fire.

Think of it like this: remember those Magic Eye posters? You'd stare and stare, seeing nothing but a jumbled mess. Then, bam! The 3D image pops out, and you can't unsee it. Climate change is kind of like that. We've been staring at the jumbled mess of extreme weather, melting ice caps, and heatwaves that feel like you're living inside a toaster oven. And suddenly, bam! – the connection between our tailpipes and that melting glacier becomes crystal clear.

We're all starting to crave a little less "toaster oven" and a little more "fresh air."

This isn't just touchy-feely stuff, either. The brainiacs over at journals like Social Psychology Quarterly are digging into the why behind this shift. They're finding that it's a beautiful mix of things: peer pressure (your neighbor got one, and suddenly you want one), a growing sense that saving the planet is actually pretty cool, and the realization that EVs aren't just good for the planet; they're also kinda awesome to drive (hello, instant torque!).

It's like that moment in high school when the "uncool" kid suddenly becomes a trendsetter. Everyone wants to be part of the movement.

And let's give credit where credit is due: the car companies finally got the memo. Volkswagen, for example, isn't just selling metal and batteries; they're selling a feeling. They're painting a picture of a future where your morning commute doesn't involve a symphony of honking and the smell of burnt gasoline. They're selling quiet, zippy rides, and a sense that you're part of something bigger – a solution, not just a consumer. It is like letting the customer take the reins, and take the car for a spin.
They're not yelling about polar bears; they're showing you how good it feels to be part of the change.

Of course, it's not all sunshine and roses. We still need more charging stations (imagine trying to bake a cake with only half the ovens working), battery tech needs to keep improving (nobody wants to be stranded on the side of the road with a dead battery), and we need to make sure everyone can afford to join the electric revolution, not just the folks with deep pockets.

The "mindset in motion" is us, and the customer. It is all of us realizing that this is not about sacrifcing, but it is about gaining.

Electric Dreams: More Than Just a Car, It's a Vibe

Remember that future we were promised? The one that didn't rumble? It hummed. Think Blade Runner 2049. Not just the flying cars – the spinners – gliding through that perpetually damp, neon-lit megacity. It was the whole feeling of the place. Electric. Efficient. Almost eerily quiet. A world teetering on the edge, saved (or maybe doomed?) by technology. That vision, that slick, high-tech, slightly gritty aesthetic, burrowed its way into our brains. It primed us. It made us want an electric future, even before we knew how it would arrive.

And then... the Cybertruck landed. Like, really landed. Forget the marketing speak for a second. Dive into the TikTok rabbit hole. It's a wild ride of Cybertruck content: breathless first impressions, hilarious roasts, surprisingly deep dives into its engineering, and meme after meme after meme. This thing, with its angles that could cut glass, isn't just transportation. It's a statement. A rolling, stainless-steel middle finger to convention. It's both aspirational and utterly, wonderfully absurd.

Now, you might be thinking: "What's a dystopian movie got to do with a truck that looks like a low-poly video game asset?" A lot, actually. They're both feeding the same beast. The same cultural hunger. There are people that will go on and on about "semiotics" and "late-stage capitalism" (and, yeah, there's truth there). But you don't need a fancy degree to get it.

Think about Gen Z. These kids grew up mainlining the internet. They're fluent in memes, fluent in digital worlds. For them, the line between "movie magic" and "stuff I can buy" is... blurry, at best. They've seen electric vehicles as the default in their favorite games and movies. So, the Cybertruck, whether it sells a million units or ten, has already won. It's a cultural touchstone. A meme lord. A conversation piece that screams, "The future is now, and it's weird!"

And that's the real magic trick here. It's not just about shifting units. It's about shifting minds. The buzz around EVs, amplified by Hollywood's sleek visions and the internet's chaotic energy, is creating its own gravity. The more EVs are seen as cool, as the future, the more people – especially the younger crowd – will want in. It's a prophecy fulfilling itself, powered by cultural forces that are way more potent than any ad budget.

So, yeah, the whisper-quiet spinner in Blade Runner and the unapologetically bold Cybertruck are worlds apart. But they're both humming the same tune. It's the song of an electric future. A future Gen Z isn't just ready for – they're demanding it. A future where technology doesn't just work; it performs. It's on our screens. It's in our feeds. And, increasingly, it's on our streets. Get ready. The electric dream is no longer just a dream. It's trending.

Electric Avenues, or Dusty Trails? The EV Dream Needs a Reality Check

The electric vehicle (EV) revolution is humming along, whispering promises of cleaner air and a greener planet. But hold on a second. Are we all invited to this electric party, or are some of us getting left at the curb? In South Africa, that question hits particularly hard, because the road to an electric future could easily widen the already massive gap between the haves and have-nots.

Imagine this: You're cruising in your sleek EV in Sandton, Johannesburg. Charging stations? No problem! They're practically sprouting from the pavement – at the mall, your gym, even that trendy new coffee shop. Life's electric, right?

Now, picture a different scene. You're in a rural village in the Eastern Cape. Your neighbor just got a donkey cart, and you're still saving up for a decent bicycle. The nearest EV charging station? Forget about it. It's probably a mirage shimmering on a distant horizon, a day's journey away, assuming the electricity even works that day. That, my friends, is the stark reality of South Africa's "grid gap" – and it's a chasm, not a gap.

This isn't just about maps and wires; it's about people. It's about whether Gogo (Grandmother) in the village can get to the clinic easily, or whether a young entrepreneur can transport goods to market without spending a fortune on petrol. It's about basic fairness. Because let's be real: access to reliable power, the very fuel of the EV revolution, is wildly different depending on where you live. Thanks to the painful legacy of apartheid, inequality is practically baked into the South African landscape. And if we're not careful, the EV rollout could end up being another chapter in that same old story.

We can't let the EV transition become the new "digital divide," just with wheels instead of Wi-Fi. So, what's the solution? How do we make sure this electric dream doesn't turn into a nightmare for those already struggling?

Let's peek over at Kenya. They're wrestling with similar issues, but they're also cooking up some seriously cool solutions. Picture this: solar-powered charging kiosks popping up in villages that have never had reliable electricity. Think electric boda-bodas (motorcycle taxis) – the lifeblood of rural transport – becoming affordable and accessible to local entrepreneurs. These aren't just charity projects; they're smart, sustainable business models, showing that going green can also mean making a living.

The key takeaway? We can't just slap charging stations down randomly and call it a day. We need to listen to the people on the ground. Does a farmer need a super-fast charger on the highway, or a slower, steadier one powered by a local solar panel setup? The second. It's like giving someone a fishing rod instead of just a fish.

Subtopic 4: From Wrenches to Wires: Germany's EV Gamble and the Rebirth of the Auto Mechanic

Forget the grease-stained overalls and the roar of a finely-tuned engine. In Germany, the soundtrack of the automotive future is a subtle electric hum, and the tools of the trade are increasingly digital. We're not just talking about a new model year; we're talking about a complete Berufsbild overhaul – a reimagining of the very identity of the auto mechanic. Germany, a land where the car is practically a cultural icon, is betting big on electric vehicles, and that bet hinges on a massive, almost audacious, human transformation.

The goal? 100,000 newly minted EV technicians by 2025. Not just "trained" technicians, but reborn ones. Imagine Günther, a 55-year-old mechanic who could diagnose a sputtering engine with a single, practiced listen. For decades, his hands knew the feel of pistons and carburetors like a sculptor knows clay. Now, Günther, and thousands like him, are trading their wrenches for wiring diagrams, their familiarity with combustion for the silent power of high-voltage batteries.

This isn't a weekend workshop; it's a full-blown technological baptism. As the Education & Training journal points out, it's not just about learning new things; it's about unlearning old ones. It's like asking a master chef, famed for their perfectly roasted chicken, to suddenly become a sushi master. The underlying principles of flavor might be similar, but the techniques, the tools, the very feel of the work are profoundly different.

But Germany, with its characteristic blend of engineering pragmatism and long-term vision, isn't leaving this to chance. They're not just throwing textbooks at these mechanics; they're throwing them into the deep end – but with life preservers made of cutting-edge training. Think of Toyota's academies in Germany, not as classrooms, but as EV dojos. Here, seasoned mechanics and fresh-faced apprentices alike grapple with the intricacies of electric powertrains, not on paper, but in real, humming, (sometimes sparking!) reality.

These aren't theoretical exercises. Imagine the quiet concentration as a technician, guided by a virtual reality overlay, traces the flow of electrons through a complex inverter. Picture the "aha!" moment when a seemingly insurmountable software glitch is finally cracked, the solution found not in brute force, but in lines of code. This is learning by doing, by failing, and by ultimately mastering a new technological language.

This isn't just about keeping Germany's auto industry humming; it's about securing its future. A skilled EV workforce isn't just a nice-to-have; it's a national imperative, a crucial piece in the puzzle of maintaining global leadership in a rapidly electrifying world. It's about economic security, yes, but it's also about something deeper: the pride of a nation built on automotive excellence.

The ripple effects will be felt far beyond the factory floor. Cheaper EV maintenance will fuel adoption. A workforce fluent in EV technology will be a breeding ground for innovation. But the most compelling story here isn't about economic indicators; it's about people like Günther. It's about the willingness to embrace a future that looks nothing like the past, to reinvent oneself, to prove that even the most ingrained skills can be rewired, that learning is a lifelong journey, not a destination. It's a story of German Zukunftsmusik – the music of the future – and it's powered not just by electricity, but by human resilience and the enduring spirit of adaptation.

Trailblazers: Landmark EV and SSB Stories

Tesla's Battery Empire: Where Electrons, Not Gasoline, Rule

Forget oil barons; the new royalty reigns in the realm of electrons, and Tesla is building its kingdom, brick by lithium-ion brick. At the heart of this electric empire? The Gigafactory – not just a factory, but a statement. It's a bold declaration that the age of gasoline is fading, and the electric future is roaring to life.

We're not talking about tinkering around the edges here. We're talking about a full-blown revolution, powered by a mind-boggling number: 50 GWh. That's Gigawatt-hours, and it's the kind of energy that makes you sit up and take notice. Imagine a single factory – like the colossal one humming in the Nevada desert, or the rapidly growing behemoth in Shanghai – churning out enough battery power to juice up hundreds of thousands of electric cars every single year.

But it's not just about bragging rights. It's about making electric cars affordable. Think of it this way: baking a single loaf of bread at home is a lovely, artisanal experience, but it's not exactly cheap. Now picture a massive bakery, pumping out thousands of loaves. Suddenly, the cost per loaf plummets. That's the Gigafactory effect.

Tesla's secret weapon is scale. They're not just building batteries; they're building a battery-making machine. By producing on this epic scale, they're slashing costs – some experts, the ones who publish in brainy journals like Energy Economics, say by as much as 30% for the whole car. That's not just a little price drop; that's a game-changer. It's like suddenly making gourmet meals as affordable as fast food.

The Shanghai Gigafactory is like Tesla's training ground. It is the testing facility for its technology, and a chance to learn. Every battery cell that rolls off the line, every robot arm that whirs into action, it's all feeding data back into the system, making the whole process smoother, cheaper, better. It's like a living, breathing organism, constantly evolving and improving.

And this isn't just good news for Tesla drivers. This battery bonanza is sending shockwaves through the entire car industry. Competitors are scrambling to catch up, driving a frenzy of innovation in everything from digging up lithium to figuring out how to recycle old batteries. It's a technological arms race, but instead of weapons, they're building a cleaner, greener future.

The Gigafactory isn't just about making cars; it's about controlling the whole game. Tesla's not just buying batteries; they're building them, from the ground up. This gives them incredible power – not just electrical power, but the power to shape their own destiny, to control their supply chain, to dictate the pace of innovation.

That 50 GWh number? It's more than a statistic. It's a battle cry. It's a symbol of a world where electric cars aren't a luxury, but the norm. It's the sound of the future, and it's getting louder every day. The kindgom is being built, and the fuel is not gasoline. It's electric.

Toyota's Silent Revolution: The 1,000km Whisper

Toyota doesn't do fireworks. They do fireflies. Tiny, persistent glimmers of light, guiding them through the long night. Their solid-state battery (SSB) project isn't a flashy press conference; it's a decade-long meditation, a quiet revolution brewing in the heart of Japan. Forget the hype – this is about hush.

Imagine a team of engineers, not in sterile white coats, but with sleeves rolled up, fueled by ramen and the unwavering belief that they're on the cusp of something monumental. For ten years, they've been wrestling with the atomic world, coaxing electrons to dance a new kind of jig. This isn't a sprint; it's a kintsugi project – carefully piecing together fragments of knowledge, mending setbacks with gold-dusted determination.

Their goal? To banish "range anxiety" – that little gremlin that sits on the shoulder of every EV driver – to the realm of myth. They're aiming for a 1,000-kilometer range. That's not just a number; it's freedom. It's the open road, unspooling like a ribbon, inviting you to explore without the constant, nagging worry of finding the next charging station. It's the difference between a weekend jaunt and a true adventure.

The secret sauce? Think of it like this: your typical lithium-ion battery is like a waterbed – sloshy, potentially leaky, and a bit temperamental. Toyota's SSB is like a beautifully crafted piece of jade – solid, stable, and packed with hidden power. They've replaced the liquid electrolyte with a solid one, a seemingly small change that unlocks a whole new world of possibilities: more energy, less weight, and, crucially, way less chance of going up in flames.

But this isn't just theory confined to the pages of Automotive Innovation. This is real. Picture the bustling streets of Tokyo, not as a backdrop, but as a crucible. Prototype vehicles, anonymous and unassuming, are navigating the urban chaos. They're enduring the stop-start ballet of rush hour, the sudden bursts of speed, the sweltering summer heat and the biting winter chill. Every pothole, every traffic jam, every sudden brake is a data point, feeding back into the lab, whispering secrets to the engineers.

Toyota's not one for boasting. They're the strong, silent type. They let the results do the talking. There's a quiet confidence emanating from their research facilities, a sense that they're not just building a better battery; they're building a future.

It's not just a year; it's a destination. It's the culmination of a decade's worth of sweat, science, and unwavering belief. If – and it's still an "if," because the world of battery tech is notoriously fickle – Toyota pulls this off, it won't just be a headline; it will be a tectonic shift. It will be the moment when electric vehicles truly come of age, shedding their limitations and becoming the undisputed kings of the road. It will be proof that sometimes, the quietest revolutions are the most powerful. The whisper will become the roar. The end.

Norway: Where Electric Dreams Are Made of... Fjords and Free Parking!

Norway isn't just talking about an electric car future. They're living it, and frankly, putting the rest of us to shame. Imagine a land where 95% – ninety-five freakin' percent! – of new car sales are electric. It's like a parallel universe where gas stations are dusty relics of a bygone era, and the air smells suspiciously like...well, clean air. This isn't science fiction; it's Norway, 2023.

How did they achieve this automotive utopia? It wasn't magic, although the sheer scale of it feels that way. It was a decades-long, carefully orchestrated symphony of policy – a policy so bold, so generous, it makes you want to pack your bags and learn Norwegian (Bokmål, preferably).

Forget gentle encouragement. Norway went full-on Viking, wielding a mighty axe of incentives. The biggest, shiniest weapon in their arsenal? Free parking and zero tolls for EV owners.

Picture this: You're in Oslo, a city where finding parking is usually a competitive sport, a gladiatorial battle for a sliver of asphalt. But you? You're cruising in your sleek, silent EV, gliding past the frustrated masses circling like sharks. You pull into a prime spot, for free, feeling a surge of smug satisfaction that's almost as electrifying as your car. And those toll booths? They're just scenic overlooks for you, my friend. You're basically the VIP of the roadways, except instead of paying a premium, you're saving a fortune.

It is not a whim, it is supported by science. A study in Transport Policy – those clever folks know their stuff – confirmed that these perks, along with hefty tax breaks on EV purchases, were like catnip to Norwegian car buyers. It's simple economics, really. Make something cheaper and more convenient, and people will flock to it.

But Norway didn't just dangle carrots. They also wielded a stick (a sustainably-sourced, ethically-harvested stick, of course). Taxes on gas-guzzlers went up, up, up, making them about as appealing as a week-old lutefisk. It's the classic "push-pull" – make the good stuff irresistible, and the bad stuff...well, less so.

However, it's more than just kroner and øre. There's a palpable feeling in Norway, a sense of collective purpose. Norwegians haven't just embraced EVs because they're financially savvy (although they are); they've embraced them because they care about the planet. Owning an EV has become a badge of honor, a symbol of their commitment to a greener future. It's like wearing a "Save the Fjords" t-shirt, only way cooler.

The outcome? A vibrant, buzzing EV ecosystem. Charging stations are as common as...well, as common as stunningly beautiful landscapes in Norway. "Range anxiety"? That's so 2010. And the used EV market is booming, making electric mobility accessible to even more people.

Yes, Norway has advantages. They're swimming in oil money (which they're ironically using to get off oil – talk about a plot twist!). But the core message is universal: bold policies, consistent commitment, and a genuine belief in a better future can work wonders.

Norway's "Electric Nirvana" isn't a fantasy. It's a real-life, working model of how to build a sustainable transportation future. It's a testament to the power of smart policy, combined with a population that's willing to embrace change. It's a message to the world: the electric future isn't some distant dream; it's here, it's now, and it's parked for free in a prime spot in Oslo. And it is the only solution, for everyone.

California's Electric Dream: Dragging Detroit (and the World) Along for the Ride

California's not just asking the auto industry to build electric cars. It's more like they've grabbed Detroit by the lapels, looked them dead in the eye, and said, "Electric is the future, get on board, or get left behind." We're not talking about polite suggestions here; we're talking about a full-throttle, regulatory earthquake shaking the foundations of how we drive. The big, bold headline? No new gas-guzzlers sold in California after 2035. Period.

This isn't some hazy, far-off fantasy. This is happening. And the tremors are echoing far beyond the Hollywood Hills.

The heart of this electric revolution is something called Advanced Clean Cars II (ACC II), but let's call it what it really is: California's "ICE Age Extinction Event." It's a meticulously crafted set of rules that progressively tighten the screws on gas-powered cars, year after year, like a boa constrictor slowly squeezing the life out of the internal combustion engine.

Imagine a relentless ratchet, clicking tighter and tighter. Automakers can't just dip their toes in the EV pool; they have to dive in headfirst, now. They have to reinvent their factories, retrain their workforce, and reimagine their entire business model, or they'll be shut out of one of the biggest car-buying playgrounds on the planet. And California isn't just playing with percentages; they're chasing real-world change, like getting 1.5 million EVs buzzing around the state by 2025 (and they're surprisingly close).

But here's where California's real superpower comes in: the legendary "California Effect." It's like the automotive equivalent of being the cool kid in high school. Because of some legal quirks (waivers under the Clean Air Act), California gets to set its own, tougher rules for car emissions. And when California says "jump," a bunch of other states ask "how high?" They adopt California's standards, creating a massive domino effect that essentially reshapes a huge chunk of the US car market.

The proof is in the pudding. Look at General Motors. Once the poster child for American muscle and gas-guzzling behemoths, GM is now singing the praises of an all-electric future. They're not doing it because they suddenly sprouted a conscience; they're doing it because California's holding a regulatory gun to their head (a very eco-friendly, zero-emission gun, of course). Imagine the sheer chaos, the mountains of cash, the mountains of effort it takes for a company like GM to completely transform itself. That's the sheer force of California's will.

The brainy folks over at the Environmental Law Review are having a field day dissecting all this. They're arguing about whether California's gone too far, whether it's a brilliant stroke of genius, or if the cost to change will be too much for people. Some say it's an overreach, others hail it as the bold leadership we desperately need to fight climate change. One thing's for sure: the debate is hotter than a California wildfire, and it's far from over.

There will be lawsuits. There will be whining from car companies. But the electric train has undeniably left the station, and it's picking up speed.

California's electric dream isn't just about breathing easier in Los Angeles. It's about forcing a global revolution in how we move. It's a grand, audacious experiment, and the whole world is watching with bated breath to see if this regulatory muscle can truly build a cleaner, greener future for us all, and possibly, save us from ourselves.

Horizons Ahead: Challenges and Promises

The Solid-State Battery Dream: Held Hostage by Tiny Cracks

Imagine a world where your electric car can zip you across the country on a single charge, topping up in the time it takes to grab a coffee. No more range anxiety, no more overnight charging marathons. That's the shimmering promise of solid-state batteries (SSBs) – the next-generation powerhouses poised to revolutionize everything from EVs to personal electronics. They're the cool kids on the battery block, promising to be safer, more powerful, and faster-charging than the lithium-ion batteries we use today.

But there's a villain in this story, a microscopic menace that's holding back the SSB revolution: anode cracking. It's the tiny, frustrating flaw that's making even the brightest battery scientists pull their hair out.

Let's break it down. Think of a battery like a seesaw. On one side, you have the cathode (the "positive" side), and on the other, the anode (the "negative" side). Lithium ions, the tiny workhorses of the battery, swing back and forth between these two sides during charging and discharging.

In today's lithium-ion batteries, the anode is usually made of graphite – think of it as a reliable, slightly boring but dependable friend. It handles the lithium-ion traffic pretty well, expanding and contracting a little, but generally keeping its cool.

But SSBs are aiming for something much more ambitious. They want to use materials like lithium metal or silicon for the anode. Why? Because these materials are like lithium-ion magnets – they can hold a ton more energy, meaning a drastically longer battery life. Imagine upgrading from a sedan to a spaceship.

The problem? These high-capacity anode materials are divas. Lithium metal, in particular, is like a balloon animal at a kid's birthday party – it expands and contracts dramatically as lithium ions come and go. Silicon is similar, though slightly less extreme. This constant swelling and shrinking puts enormous stress on the anode.

Imagine squeezing and stretching a piece of clay over and over again. Eventually, what happens? It cracks. And that's exactly what happens to these super-powered anodes. Tiny cracks appear, then grow, and before you know it, your dream battery is performing like a dud. Performance plummets, lifespan shortens, and the whole thing can even fail catastrophically. It's like building a magnificent skyscraper on a foundation of sand.

It's a huge headache. You can have the most brilliant solid electrolyte (the "road" the lithium ions travel on in an SSB) and the most perfect cathode, but if your anode is turning into dust, your SSB is going nowhere fast.

So, what's the solution? It's a bit like microscopic architecture meets materials science wizardry.

One promising approach is "doping" silicon, a technique that's being explored at places like Stanford. Think of it like adding steel rebar to concrete – introducing tiny amounts of other elements into the silicon structure to make it stronger and more resilient. These "dopants" are like tiny stress relievers, helping the silicon handle the expansion and contraction without falling apart.

There's serious science backing this up, published in journals like Nano Letters, and companies like Amprius Technologies are putting these ideas to the test with silicon nanowire anodes. They're battling the real-world challenges of scaling up these technologies, learning from every crack and failure.

It's like a high-stakes, microscopic game of Tetris. Scientists are meticulously tweaking the anode's structure, experimenting with different "dopant" ingredients, different shapes (like nanowires), and different manufacturing recipes, all trying to find the sweet spot – an anode that's both incredibly powerful and incredibly durable.

The anode cracking problem is a major roadblock, no doubt about it. But it's not a dead end. The sheer amount of brainpower and innovation being thrown at this problem – from university labs to corporate R&D departments – is inspiring. We might not be plugging our cars into SSB chargers next week, but every successful experiment, every tiny improvement in anode stability, brings us closer to that energy storage revolution. The future is being built, one crack-resistant anode at a time. And it's going to be electrifying.

Bolivia's Lithium Dreams: A High-Stakes Game of Battery Bingo

Imagine this: a shimmering, alien landscape. The Salar de Uyuni in Bolivia. It's not just salt; it's the future, spread out like a giant, blindingly white tablecloth. Underneath? The "white gold" – lithium. Enough to make every electric car on Earth hum, and potentially, to shift the balance of global power.

Bolivia, bless its heart, has been sitting on this treasure chest for ages, like a kid with the winning lottery ticket but no idea how to cash it. Extracting lithium from this otherworldly brine is tricky. It's like trying to separate a single grain of sugar from a swimming pool. And Bolivia hasn't had the right equipment, the right know-how, or, frankly, the political stability to pull it off solo. Past attempts at partnerships? Let's just say they've ended up more like messy divorces than happy marriages.

Enter China, the world's EV-making machine. China needs lithium like a marathon runner needs water – desperately, and in vast quantities. They're not just thirsty; they're strategic. They've realized something crucial: It's not enough to just dig up the lithium. You need to refine it. It's like having a mountain of coffee beans but no coffee maker. Useless.

China has quietly, patiently, become the world's master coffee maker of lithium. They've built the refineries, the processing plants, the whole shebang. Even if Bolivia, or Chile, or Argentina (the "Lithium Triangle," picture a South American Bermuda Triangle, but instead of disappearing ships, it's all about disappearing battery independence) manages to yank tons of lithium out of the ground, it often ends up on a boat to China.

This is where it gets a little… uneasy. Imagine a game of Bingo. Bolivia has the lucky number (the lithium), but China owns the Bingo hall and the machine that calls the numbers (the refining process). Who do you think is going to yell "Bingo!" first, and most often?

The European Union is watching this game unfold with a growing sense of, shall we say, nervousness. They're all in on electric cars, but they've woken up to the fact that they're basically begging other countries for the keys to the kingdom. Their "diversification push" is basically a frantic search for other Bingo halls, other number-calling machines, anything to avoid being completely at the mercy of a single supplier. They're exploring deals in Australia, whispering sweet nothings to African nations, even digging around in their own backyards, hoping to find a little lithium sparkle.

This isn't a good-versus-evil story. Bolivia wants to benefit from its natural riches, and who can blame them? China is pursuing its national interests, just like any other major power. But the imbalance is the real story. It is a story that shows the changing of the guard, in real time.

It is a global game of resource chess, played out on a scale that's almost too big to grasp. It's about who controls the battery juice that will power our future – our cars, our homes, maybe even our lives. And right now, the game is far from over, but one thing's for sure: the players are getting serious, and the stakes are higher than ever. This isn't just about batteries; it's about power, plain and simple.

The Great Car-Sharing, Self-Driving Tango: A Greener Groove for Our Cities?

Okay, folks, buckle up (or maybe unbuckle, since you won't be driving!), because the future of getting around is shaping up to be less Mad Max and more… synchronized swimming. We're not just talking about robot cars anymore. We're talking about a whole new rhythm for city life, a beautiful, coordinated dance between self-driving technology and the idea that sharing is caring (especially when it comes to our planet). This is the dawn of "Mobility Fusion," and it might just be the breath of fresh air our cities desperately need.

Imagine this: Waymo, the folks who are basically teaching cars to be smarter than your average teenager, aren't just building robotaxis. They're building fleets of them. Think of it like a perfectly choreographed ballet of electric vehicles, all powered by those super-powered Solid-State Batteries (SSBs) – the ones that charge faster than you can say "range anxiety" and last longer than your last questionable online purchase. The dream? These Waymo-SSB super-teams could slash city emissions by a jaw-dropping 20%. That's not just a statistic; that's kids breathing easier, skies looking bluer, and maybe even a few less coughs and sniffles all around.

So, how does this magic trick work? It's all about efficiency, baby! First, electric vehicles are inherently cleaner – no more tailpipes pumping out yuck. Then, add the superpowers of SSBs. These batteries let Waymo's vehicles stay on the road longer, like the Energizer Bunny of transportation, serving more riders and spending less time glued to a charger. It's like a constant game of passenger Tetris, maximizing every ride and minimizing wasted time.

But the real secret sauce is the sharing. Let's be honest, most of our cars are basically expensive lawn ornaments for most of the day. They sit there, lonely and unused, while we're at work, watching cat videos, or, you know, living our lives. Shared autonomous fleets are the opposite. They're like the ultimate party animals of the car world – always on the go, picking people up, dropping them off, and then zipping off to the next adventure. This means we need way fewer cars overall, which is a direct win for Mother Earth. It's like a carpool on steroids, fueled by algorithms and a desire for a cleaner planet.

This isn't some futuristic fantasy, either. The brainy folks in lab coats are churning out research papers in places like Transportation Research, proving this isn't just hot air. And we're not just talking about computer models; real-world data is starting to roll in. Take Phoenix, Arizona, for example. It's like Waymo's personal playground, where they're testing this whole self-driving, car-sharing shebang in a real-world setting. They're learning the ropes, figuring out the kinks, and getting a sneak peek at how this mobility revolution might unfold.

Phoenix is more than a test zone. Phoenix is proving the concept. The rider surveys, the use patterns, and – most importantly – the data on how much cleaner the air is getting – that's gold dust. It's the evidence that this isn't just a cool idea; it's a real, working solution.

Sure, there are still some bumps in the road (pun intended!). We need to get people comfortable with the idea of robot drivers, build the infrastructure to support all these electric vehicles, and keep making the technology even smarter. But the prize – a future where getting around is cleaner, easier, and more accessible – is too good to pass up. That 20% emission reduction? That's not just a number; it's a promise. A promise of cities that are a little less choked, a little more vibrant, and a whole lot more sustainable. It's a future worth driving – or, rather, being driven – towards.

The Battery's Secret Second Life: From Phone Zombie to Eco-Warrior

Remember that heart-stopping moment when your phone blinks that dreaded 1% battery warning? We've all been there, that modern-day equivalent of a near-death experience. We treat our batteries like tireless workhorses, demanding constant power, and then... well, traditionally, we tossed them aside like yesterday's news. But what if I told you your phone battery, even after it's "dead," has a secret second life? A much cooler one than you might imagine.

Forget the graveyard of forgotten electronics. We're entering the era of the battery renaissance. Think of it less like a linear path to the landfill and more like a phoenix rising from the ashes – only instead of ashes, it's a meticulously sorted pile of valuable metals. And no, the pheonix won't be burning anything.

This isn't your grandpa's recycling. We're not talking about some feel-good, token effort that barely makes a dent. We're talking about a full-blown, 90% material recovery revolution, especially for those fancy new Solid-State Batteries (SSBs) that are poised to power the future.

Picture this: instead of a fiery furnace (which, let's be honest, sounds a bit like a medieval torture device for batteries), imagine a high-tech spa treatment. That's essentially what Duesenfeld, a German company with a name that sounds like a Bond villain but is actually an eco-hero, is doing.

They're like the master chefs of battery recycling. Instead of throwing the whole battery into a blender (ouch!), they carefully disassemble it in a nitrogen-filled room (no fiery explosions here!). It's like a delicate surgical operation, separating the lithium, nickel, cobalt, and manganese – the battery's "superfood" ingredients – with precision. They then use special, eco-friendly "sauces" (solvents) to extract these precious metals in a pure, ready-to-use form.

These recovered goodies aren't destined for some dusty shelf. They're reborn, ready to power new batteries. It's the ultimate closed-loop system, the circle of (battery) life, Simba! And it's a far cry from the old days of digging up mountains and relying on far-flung corners of the world for these resources.

This is a big deal, and not just for tree-huggers (though we salute you!). The Circular Economy journal, which is basically the Vogue of sustainability, is buzzing about it.

And big names like Renault, the car company, aren't just building electric cars; they're becoming battery lifecycle choreographers.

Renault is thinking long-term. They're setting up systems to collect old batteries and ensure those precious metals get a second (and third, and fourth...) chance at powering our lives. They're basically building a battery retirement community where the residents get to be useful again, instead of ending up in a landfill.

Think of it this way: the volatile global supply chain for battery materials is like a rollercoaster – unpredictable and potentially scary. Recycling, on the other hand, is like building a stable, local train line. It's more reliable, more resilient, and keeps things moving smoothly.

We're still at the dawn of this battery revolution. But the wheels are turning (pun intended!). With innovative geniuses like Duesenfeld and forward-thinking companies like Renault leading the charge, the future of batteries is looking bright – and surprisingly circular. Your "dead" phone battery might just be the key to a greener, more sustainable tomorrow. It's not just about extending battery life; it's about giving it a whole new afterlife. And that's a story worth plugging into.

The Road Forward: Vision and Victory

The Hum of Change: A Quiet Revolution on Wheels (and Beyond)

We're living in a movie, aren't we? The one where the climate's gone a bit haywire, less "gentle breeze" and more "battering ram" at our collective front door. It's scary, no doubt. But hold on, because amidst the drama, there's a subplot brewing, a good one, and it's got a surprisingly quiet engine. It's the story of two unsung heroes: Electric Vehicles (EVs) and Solid-State Batteries (SSBs). They're not just shiny new gadgets; they're partners in crime, fighting the good fight against climate change.

The IPCC, those folks who basically write the "science textbook" for the planet, they're usually not known for their light reading. Their reports are serious stuff. But even in their stark warnings, there's a glimmer of hope, a "how-to" guide for dodging the worst-case scenario. And guess who stars in that guide? You guessed it: EVs and SSBs. We're not talking about some futuristic fantasy here. We're talking about now.

Imagine this: by 2040, these two working together could slash a whopping 4 gigatonnes of CO2. That's like the entire European Union deciding to take a collective, carbon-neutral vacation. Permanently. It is the poluting equivalent of taking a nap for a whole year.

The most obvious application, is the one we can feel directly. We can get in a car that doesn't produce polution. Less noise, less smog... It is awesome.
But, and this is the really exciting part, it's not just about cleaner commutes. EVs and SSBs are like the ultimate

power couple. They're changing the whole game of energy.

Think of your house, basking in sunshine, solar panels soaking up the rays. That extra energy? It's not wasted. It's tucked away in a super-efficient SSB, waiting patiently for nighttime, or a cloudy day. Imagine a power grid that's less like a grumpy, coal-chugging grandpa and more like a nimble, renewable energy ninja. That's the future we're talking about.

And this isn't just some lab experiment. Look around! China's already rolling out EVs like they're going out of style, and they're betting big on SSBs. Europe's car companies are in a full-blown sprint to catch up, throwing money at research like it's confetti. And in the US, the electric buzz is getting louder, driven by everyone from policymakers to everyday folks who just want a cleaner ride. It's a global jigsaw puzzle, and all the pieces are snapping into place.

SSBs are the secret sauce, the magic ingredient. They're what make EVs go from "pretty cool" to "holy smokes, this changes everything!" They pack more power, charge faster (say goodbye to that "will I make it?" anxiety), are safer (less chance of turning into a surprise bonfire), and last longer (your wallet and the planet will thank you). They're the key that unlocks the EV's full potential, and they're speeding up our journey to a world where energy doesn't cost the Earth.

So, what have we learned? It's not just about a new kind of car. It's about a new kind of everything. It's about building a future that's cleaner, more sustainable, and frankly, a lot less stressful. It's about using technology to be, well, better humans. It's about grabbing the steering wheel of our future and pointing it in a direction that actually makes sense. And that, my friends, is a ride worth taking.

Crystal Ball Gazing: Buckle Up, Buttercup – 2032 is Gonna Be Electric!

Okay, let's ditch the stuffy futurist act for a second. Predicting the future of electric cars and batteries? It's like trying to herd caffeinated squirrels – chaotic, unpredictable, but undeniably exciting. But I've been digging into the data, and folks, the next ten years are shaping up to be a wild, electrifying ride.

The big, bold headline? We're staring down the barrel of 70% EV adoption by 2032. I'm not talking about some niche market for folks who hug trees for fun (no offense to tree-huggers!). This is your neighbor, your grandma, your dentist potentially going electric. Remember when seeing a Tesla felt like spotting a unicorn? Now they're practically multiplying like rabbits. That's not a blip; it's a full-blown stampede.

And I'm not just pulling this out of my... well, you know. We're talking about smart folks at places like Nature Energy, crunching the numbers, tracking the trends. They're seeing the same writing on the wall: cheaper batteries, governments getting serious about emissions, and people finally realizing that electric cars aren't just "good for the planet," they're also awesome to drive. It's a perfect storm, brewing up a revolution on wheels.

But here's where things get really spicy: Solid-State Batteries (SSBs) – the mythical beasts of the battery world – are predicted to hit cost parity with regular lithium-ion batteries by 2032. Translation? The holy grail of EV tech is about to become... affordable.

Let me break it down for the non-battery geeks (I see you!). SSBs are like the superheroes of batteries. They pack more punch (longer range), recharge faster than you can say "supercharger" (we're talking minutes, people!), and are way less likely to go all "spontaneous combustion" on you.

Right now, SSBs are priced like they're made of solid gold, sprinkled with unicorn tears. That's why you don't see them in every electric scooter and golf cart. But think back... Remember when a flat-screen TV cost more than a used car? Or when cell phones were the size of bricks and only for Wall Street bigwigs? Technology has a funny way of starting out ridiculously expensive and then becoming so cheap everyone has one (or two).

That's the SSB story, unfolding in real-time. Labs are buzzing, billions are being poured in, and companies are in a full-on sprint to figure out how to mass-produce these miracle batteries without breaking the bank. And when they do? Kaboom.

It is a seismic shift in how we view the technology. Imagine: Remember when only fancy-schmancy luxury cars had those cool features like GPS and parking assist? Now, even your basic hatchback probably has them. That's the power of affordability and desirability colliding.

So, grab your helmet, adjust your mirrors, and get ready for a decade of pure automotive adrenaline. We're not just talking about a few tweaks here and there; we're talking about a complete reimagining of how we get from point A to point B. 70% EVs and affordable super-batteries? It's not a pipe dream; it's looking more and more like our future reality. And honestly? It's going to be a blast.

Rallying Cry: Power Up the Planet – One Trillion Strong, by 2030!

Listen up. We're not talking about some far-off, sci-fi future anymore. We're talking about right now. The electric vehicle revolution is rumbling to life, but it needs a serious jolt of lightning – we need to crank this thing up to eleven. This isn't just about swapping gas guzzlers for sleek EVs; it's about rewiring how the entire world moves. And that? That takes guts, vision, and a whole lot of collaborative muscle.

That's why we're throwing down the gauntlet: $1 trillion invested globally in EV infrastructure by 2030.

This isn't some pie-in-the-sky, wishful-thinking number. We've crunched the data, stared down the challenge, and this is what it really takes to ignite this revolution. We're not just talking about a few chargers sprinkled around fancy city centers. We're talking about blanketing the planet – from bustling megacities to sleepy rural towns, from superhighways to the backroads of developing nations. We need a grid that can handle the surge, batteries that go the distance (and then some!), and smart ways to recycle them when they're done. We need to make sure that the folks who built our cars for generations have a place in this new, electric world – a just transition is non-negotiable.

The World Bank? They've been shouting this from the rooftops. Their reports are practically begging for massive investment to unlock the green transportation future we all crave. And those promises made at COP28? They're beautiful words, but words need action, concrete beneath their feet. We need to build the future, not just talk about it.

But here's the kicker: this isn't about governments going it alone. This is about a symphony of action. Governments, you're the conductors – set the stage with smart policies, tempting incentives, and cut through the red tape. Private sector, you're the orchestra – bring the innovation, the investment, the sheer drive to make this happen. Think epic partnerships: automakers jamming with energy companies, tech wizards collaborating with local communities.

Picture this: you're cruising down the open road, the wind in your hair (or helmet!), and range anxiety? That's ancient history. Charging up is as easy as grabbing a coffee. The air is crisp and clean, and the only sound is the whisper-quiet hum of your electric ride. This isn't a fantasy; it's within our grasp, but it takes all of us pulling together.

So, here's the call to arms:

    Governments: Be bold! Set those ambitious targets, and make them real.
    Businesses: See the future? It's electric, and it's profitable. Invest now, and reap the rewards.
    Innovators: Keep pushing the boundaries! Blow our minds with the next generation of battery tech.
    Everyone: Demand better. Choose electric. Be the change.

This isn't just about wires and chargers; it's about powering a better future. A future where getting around is sustainable, affordable, and fair for everyone. It's about creating jobs, sparking incredible innovation, and staring down climate change with a defiant grin.

One trillion dollars by 2030. It's a huge number, sure, but the payoff? A healthier planet, a booming economy, and a future we can all be proud of – that's priceless.

Let's not just make this happen; let's electrify it! Let's unleash our collective creativity, our passion, our unwavering determination. We, as humans, are capable of incredible things when we work together.

Let that number – one trillion – become more than a goal. Let it be a battle cry, a symbol of hope, a movement that changes the world. One plug-in at a time.

The Zero-Emission Symphony: An Audacious Dream

Let's be honest, the air's getting thick, isn't it? We're choking on the fumes of yesterday's solutions. Fossil fuels? They're the dinosaurs of the energy world, and we all know what happened to the dinosaurs. It's not a secret; it's a screaming alarm bell. But here's the thing: knowing the problem is only the first chord in a much grander symphony. It's the doing, the relentless creating, that writes the rest of the music.

We're not talking about a quick jog around the block here. This zero-emission thing? It's not even a marathon. It's more like an eternal relay race, a cosmic hand-off of the creative baton. Each generation grabs it, sprints like hell, and passes it on to the next, fueled by a crazy, beautiful, almost foolish optimism.

And this isn't some fluffy, feel-good mantra. This is grounded in the gritty reality of Innovation Studies. Think of it as the science of "what if?". It's about dissecting how wild ideas sprout, how they kick the status quo in the teeth, and, most importantly, how they fix the messes we've made. It's about taking "impossible" and smashing it into a million manageable pieces.

Take Elon Musk, for example. Love him or hate him, the guy's got guts. He didn't just want to tweak the car; he wanted to reinvent the whole damn road. Electric vehicles went from nerdy novelties to serious contenders. Reusable rockets? That's straight-up sci-fi becoming Tuesday afternoon. Solar panels and batteries are getting better, faster, stronger, like some kind of technological superhero training montage. This isn't hocus pocus; it's the raw power of human brains refusing to accept "no" for an answer.

Musk's vibe, even if it sometimes veers into "mad scientist" territory, is the essence of what we need. It's not about making a slightly less polluting gas guzzler. It's about making the gas guzzler a museum piece. That kind of radical, tear-it-all-down-and-build-it-better thinking is the engine of this revolution.

But – and this is crucial – it's not a one-man show. This zero-emission future needs a global jam session. It's a symphony of innovation, where everyone's got an instrument to play: scientists, engineers, entrepreneurs, policymakers, even the artists painting murals on the sides of wind turbines.

Imagine the breakthroughs hiding just around the corner. Batteries that can hold the power of a thousand suns. Smart grids that dance with the wind and the sun. Materials so light, strong, and sustainable, they'd make nature jealous. The possibilities are as vast as the universe itself, limited only by how far we dare to dream.

And, let us be clear, there will be screw-ups. Colossal, face-plant, "back to the drawing board" screw-ups. And that's okay. Progress isn't a pristine, straight line; it's a messy, chaotic, beautiful explosion.
We need a mindset shift.

This isn't just about gadgets and gizmos. It's about cultivating a culture, a vibe of innovation. A place where experimentation is encouraged, where failure is a badge of honor (because it means you tried), and where the relentless pursuit of "better" is the only rule. It's about guts. It is about hope. It's about believing, deep down, that we can make things amazing.

That, my friends, is how we win this thing. The quest is on, and it's a quest worth every ounce of sweat, every brilliant idea, and every spectacular failure. The planet's future? It's riding on our creativity. Let's make some noise.

About Author

Azhar ul Haque Sario is a bestselling author and data scientist with a remarkable record of achievement. This Cambridge alumnus brings a wealth of knowledge to his work, holding an MBA, ACCA (Knowledge Level - FTMS College Malaysia), BBA, and several Google certifications, including specializations in Google Data Analytics, Google Digital Marketing & E-commerce, and Google Project Management.

With ten years of business experience, Azhar combines practical expertise with his impressive academic background to craft insightful books. His prolific writing has resulted in an astounding 2810 published titles, earning him the record for the maximum Kindle editions and paperback books published by an individual author in one year, awarded by Asia Books of Records in 2024.

ORCID: https://orcid.org/0009-0004-8629-830X
Azhar.sario@hotmail.co.uk
https://www.linkedin.com/in/azharulhaquesario/